T0212233

Analyzing Analytics

Synthesis Lectures on Computer Architecture

Editor
Margaret Martonosi, *Princeton University*

Synthesis Lectures on Computer Architecture publishes 50- to 100-page publications on topics pertaining to the science and art of designing, analyzing, selecting and interconnecting hardware components to create computers that meet functional, performance and cost goals. The scope will largely follow the purview of premier computer architecture conferences, such as ISCA, HPCA, MICRO, and ASPLOS.

Analyzing Analytics
Rajesh Bordawekar, Bob Blainey, and Ruchir Puri
2015

Research Infrastructures for Hardware Accelerators
Yakun Sophia Shao and David Brooks
2015

Customizable Computing
Yu-Ting Chen, Jason Cong, Michael Gill, Glenn Reinman, and Bingjun Xiao
2015

Die-stacking Architecture
Yuan Xie and Jishen Zhao
2015

Single-Instruction Multiple-Data Execution
Christopher J. Hughes and
2015

Power-Efficient Computer Architectures: Recent Advances
Magnus Själander, Margaret Martonosi, and Stefanos Kaxiras
2014

FPGA-Accelerated Simulation of Computer Systems
Hari Angepat, Derek Chiou, Eric S. Chung, and James C. Hoe
2014

A Primer on Hardware Prefetching
Babak Falsafi and Thomas F. Wenisch
2014

On-Chip Photonic Interconnects: A Computer Architect's Perspective
Christopher J. Nitta, Matthew K. Farrens, and Venkatesh Akella
2013

Optimization and Mathematical Modeling in Computer Architecture
Tony Nowatzki, Michael Ferris, Karthikeyan Sankaralingam, Cristian Estan, Nilay Vaish, and David Wood
2013

Security Basics for Computer Architects
Ruby B. Lee
2013

The Datacenter as a Computer: An Introduction to the Design of Warehouse-Scale Machines, Second edition
Luiz André Barroso, Jimmy Clidaras, and Urs Hölzle
2013

Shared-Memory Synchronization
Michael L. Scott
2013

Resilient Architecture Design for Voltage Variation
Vijay Janapa Reddi and Meeta Sharma Gupta
2013

Multithreading Architecture
Mario Nemirovsky and Dean M. Tullsen
2013

Performance Analysis and Tuning for General Purpose Graphics Processing Units (GPGPU)
Hyesoon Kim, Richard Vuduc, Sara Baghsorkhi, Jee Choi, and Wen-mei Hwu
2012

Automatic Parallelization: An Overview of Fundamental Compiler Techniques
Samuel P. Midkiff
2012

Phase Change Memory: From Devices to Systems
Moinuddin K. Qureshi, Sudhanva Gurumurthi, and Bipin Rajendran
2011

Multi-Core Cache Hierarchies
Rajeev Balasubramonian, Norman P. Jouppi, and Naveen Muralimanohar
2011

A Primer on Memory Consistency and Cache Coherence
Daniel J. Sorin, Mark D. Hill, and David A. Wood
2011

Dynamic Binary Modification: Tools, Techniques, and Applications
Kim Hazelwood
2011

Quantum Computing for Computer Architects, Second Edition
Tzvetan S. Metodi, Arvin I. Faruque, and Frederic T. Chong
2011

High Performance Datacenter Networks: Architectures, Algorithms, and Opportunities
Dennis Abts and John Kim
2011

Processor Microarchitecture: An Implementation Perspective
Antonio González, Fernando Latorre, and Grigorios Magklis
2010

Transactional Memory, 2nd edition
Tim Harris, James Larus, and Ravi Rajwar
2010

Computer Architecture Performance Evaluation Methods
Lieven Eeckhout
2010

Introduction to Reconfigurable Supercomputing
Marco Lanzagorta, Stephen Bique, and Robert Rosenberg
2009

On-Chip Networks
Natalie Enright Jerger and Li-Shiuan Peh
2009

The Memory System: You Can't Avoid It, You Can't Ignore It, You Can't Fake It
Bruce Jacob
2009

Fault Tolerant Computer Architecture
Daniel J. Sorin
2009

The Datacenter as a Computer: An Introduction to the Design of Warehouse-Scale Machines
Luiz André Barroso and Urs Hölzle
2009

Computer Architecture Techniques for Power-Efficiency
Stefanos Kaxiras and Margaret Martonosi
2008

Chip Multiprocessor Architecture: Techniques to Improve Throughput and Latency
Kunle Olukotun, Lance Hammond, and James Laudon
2007

Transactional Memory
James R. Larus and Ravi Rajwar
2006

Quantum Computing for Computer Architects
Tzvetan S. Metodi and Frederic T. Chong
2006

Analyzing Analytics

Rajesh Bordawekar, Bob Blainey, and Ruchir Puri

ISBN: 978-3-031-00621-0   paperback
ISBN: 978-3-031-01749-0   ebook

DOI 10.1007/978-3-031-01749-0

A Publication in the Springer series
SYNTHESIS LECTURES ON ADVANCES IN AUTOMOTIVE TECHNOLOGY

Lecture #35
Series Editor: Margaret Martonosi, *Princeton University*
Series ISSN
Print 1935-3235   Electronic 1935-3243

Analyzing Analytics

Rajesh Bordawekar
IBM T.J. Watson Research Center

Bob Blainey
IBM CloudLab

Ruchir Puri
IBM T.J. Watson Research Center

SYNTHESIS LECTURES ON COMPUTER ARCHITECTURE #35

ABSTRACT

This book aims to achieve the following goals: (1) to provide a high-level survey of key analytics models and algorithms without going into mathematical details; (2) to analyze the usage patterns of these models; and (3) to discuss opportunities for accelerating analytics workloads using software, hardware, and system approaches. The book first describes 14 key analytics models (exemplars) that span data mining, machine learning, and data management domains. For each analytics exemplar, we summarize its computational and runtime patterns and apply the information to evaluate parallelization and acceleration alternatives for that exemplar. Using case studies from important application domains such as deep learning, text analytics, and business intelligence (BI), we demonstrate how various software and hardware acceleration strategies are implemented in practice.

This book is intended for both experienced professionals and students who are interested in understanding core algorithms behind analytics workloads. It is designed to serve as a guide for addressing various open problems in accelerating analytics workloads, e.g., new architectural features for supporting analytics workloads, impact on programming models and runtime systems, and designing analytics systems.

KEYWORDS

analytics, parallel algorithms, hardware acceleration

Contents

1   **Introduction** 1

    1.1   Analytics: A Definition 1

    1.2   Analytics at Your Service 1

    1.3   Classification of Analytics Applications 2

       1.3.1   The Watson DeepQA System 4

       1.3.2   Functional Flow of Analytics Applications 5

    1.4   Intended Audience 9

2   **Overview of Analytics Exemplars** 11

    2.1   Exemplar Models 11

    2.2   Regression Analysis 12

    2.3   Clustering 14

    2.4   Nearest Neighbor Search 16

    2.5   Association Rule Mining 18

    2.6   Recommender Systems 21

    2.7   Support Vector Machines 23

    2.8   Neural Networks 25

    2.9   Decision Tree Learning 28

    2.10   Time Series Processing 30

    2.11   Text Analytics 33

    2.12   Monte Carlo Methods 36

    2.13   Mathematical Programming 38

    2.14   On-line Analytical Processing 41

    2.15   Graph Analytics 44

3   **Accelerating Analytics** 47

    3.1   Characterizing Analytics Exemplars 47

       3.1.1   Computational Patterns 47

       3.1.2   Runtime Characteristics 50

    3.2   Implications on Acceleration 52

       3.2.1   System Acceleration Opportunities 52

4   Accelerating Analytics in Practice: Case Studies **57**

4.1   Text Analytics 57

4.2   Deep Learning 59

4.3   Computational Finance 62

4.4   OLAP/Business Intelligence 64

4.5   Graph Analytics 66

5   Architectural Desiderata for Analytics **69**

5.1   Accelerators for Analytics Workloads 70

5.2   Bringing it all together: Building an Analytics System 74

A   Examples of Industrial Sectors and Associated Analytical Solutions **77**

Bibliography **79**

Authors' Biographies **113**

CHAPTER 1

Introduction

1.1 ANALYTICS: A DEFINITION

From streaming news updates on smartphones, to instant messages on micro-blogging sites, to posts on social network sites, we are all being overwhelmed by massive amounts of diverse data (The Economist [2010]). Access to such a large amount of diverse data can be a boon if any useful information can be extracted and applied rapidly and accurately to a problem at hand. For instance, we could contact all of our nearby friends for a dinner at a local, mutually agreeable and well-reviewed restaurant that has table availability for that night, but finding and organizing all that information can be very challenging. This process of identifying, extracting, processing, and integrating information from raw data, and then applying it to solve a problem is broadly referred to as analytics and has now become an integral part of everyday life.

1.2 ANALYTICS AT YOUR SERVICE

Tables 1.1 and 1.2 present a sample of key analytics applications from different domains, along with their functional characteristics. As these tables illustrate, many services that we take for granted and use extensively in everyday life would not be possible without analytics. For example, social networking applications such as Facebook, Twitter, and LinkedIn encode social relationships as graphs and use graph algorithms to identify hidden patterns (e.g., finding common friends). Other popular applications like Google Maps, Yelp, or FourSquare combine location and social relationship information to answer complex spatial queries (e.g., find the nearest restaurant of a particular cuisine that your friends like). Usage of analytics has substantially improved the capabilities and performance of gaming systems as demonstrated by the recent win of IBM's Watson intelligent question-answer system over human participants in the Jeopardy! challenge. The declining cost of computing and storage and the availability of such infrastructure in cloud environments has enabled organizations of any size to deploy advanced analytics and to package those analytic applications for broad usage by consumers.

While consumer analytical solutions may help us all to better organize or enrich our personal lives, the analytic process is also becoming a critical capability and competitive differentiator for modern businesses, governments and other organizations. In the current environment, organizations need to make on-time, informed decisions to succeed. Given the globalized economy, many businesses have supply chains and customers that span multiple continents. In the public sector, citizens are demanding more access to services and information than ever before. Huge improvements in communication infrastructure have resulted in widespread use of online com-

Table 1.1: Examples of well-known analytics applications

Application	Principal Goals
Google search, Bing	Web Indexing and Search
Netflix and Pandora	Video and Music Recommendation
Watson	Intelligent Question-Answer System
Telecom Churn Analysis	Analysis of Call-data Records (CDRs)
Cognos Consumer Insight (CCI)	Sentiment/Trend Analysis of BLOGS
UPS	Logistics, Transportation Routing
Amazon Web Analytics	Online Retail Management
Moodys, Fitch, S&P Analytics	Financial Credit Rating
Yelp, FourSquare	Integrated Geographical Analytics
Oracle, SAS Retail Analytics	End-to-end Retail Management
Splunk	System Management Analytics
Salesforce.com	CRM Analytics
CoreMetrics, Mint, Youtube Analytics	Web-server Workload Analytics
Expedia, Orbitz	Travel Planning and Reservation
Flickr, Twitter, Facebook and Linkedin Analytics	Social Network Analysis
Healthcare Analytics	Streaming Analytics for Intensive Care
Voice of Customer Analytics	Analyzing Customer Voice Records

merce and a boom in smart, connected mobile devices. More and more organizations are run around the clock, across multiple geographies and time zones, and those organizations are being instrumented to an unprecedented degree. This has resulted in a deluge of data that can be studied to harvest valuable information and make better decisions. In many cases, these large volumes of data must be processed rapidly in order to make timely decisions. Consequently, many organizations have employed analytics to help them decide what kind of data they should collect, how this data should be analyzed to glean key information, and how this information should be used for achieving their organizational goals. Examples of such techniques can be found in almost any sector of the economy, including financial services (Crosbie and Bohn [2003]), government (Goode [2011]), healthcare, retail (Richter et al. [2010]), manufacturing, logistics (Armacost et al. [2004]), hospitality, and eCommerce (Davenport and Harris [2007]). Appendix A presents a more exhaustive list of business analytics solutions and associated industrial sectors.

1.3 CLASSIFICATION OF ANALYTICS APPLICATIONS

As Table 1.2 illustrates, analytics applications exhibit a wide range of functional characteristics. The distinguishing feature of an analytics application is the use of mathematical formulations for modeling and processing the raw data, and for applying the extracted information. These techniques include statistical approaches, numerical linear algebraic methods, graph algorithms, relational operators, and string algorithms. In practice, an analytics application uses multiple formulations, each with unique functional and runtime characteristics (Table 1.1). Further, depending on the functional and runtime constraints, the same application can use different algorithms.

Table 1.2: Key characteristics of the analytics applications

Application	Key Functional Characteristics
Google, Bing search	Web crawling, Link analysis of the web graph, Result ranking, Indexing Multi-media data
Netflix and Pandora	Analyzing structured and unstructured data, Recommendation
Watson	Natural language processing, Processing large unstructured data, Artificial intelligence (AI) techniques for result ranking and wagering
Telecom Churn Analysis	Graph modeling of call records, Large graph dataset, Connected component identification
Cognos Consumer Insight	Processing large corpus of text documents, Extraction and Transformation, Text indexing, Entity extraction
UPS	Mathematical optimization-based solutions for transportation
Amazon Web Analytics Salesforce.com	Analysis of e-commerce transactions, Massive data sets, Real-time response, Reporting, Text search, Multi-tenant support personalization, Automated price determination, Recommendation
Moody's, Fitch, S&P	Statistical analysis of large historical data
Google Maps, Yelp	Spatial queries, Streaming and persistent data, Spatial ranking
Oracle, SAS, Amazon Retail Analytics	Analysis over large persistent and transactional data, Extraction and Transformation, Reporting, Integration with logistics, HR, CRM
Hyperic and Splunk	Text analysis of large corpus of system logs
CoreMetrics, Mint, Youtube Analytics	Website traffic/workload characterization, Massive historical data, Video annotation and search, Online advertisement/marketing
Expedia, Orbitz	Mathematic optimization-based solutions for travel industry
Flickr, Twitter, Facebook and Linkedin	Graph modeling of relations, Massive graph datasets, Graph analytics, Multi-media annotations and indexing
Healthcare Analytics	Streaming data processing, Time-series analysis
Voice of Customer Analytics	Natural language processing, Text entity extraction

While many of the applications process a large volume of data, the type of data processed varies considerably. Internet search engines process unstructured text documents as input, while retail analytics operate on structured data stored in relational databases. Some applications such as Google Maps, Yelp, or Netflix use both structured and unstructured data. The velocity of data also differs substantially across analytics applications. Search engines process read-only historical data whereas retail analytics process both historical and transactional data. Other applications, such as the monitoring of medical instruments, work exclusively on real-time or streaming data. Depending on the mathematical formulation, the volume and velocity of data and the expected I/O access patterns, the data structures and algorithms used by analytical applications vary considerably. These data structures include vectors, matrices, graphs, trees, relational tables, lists, hash-based structures, and binary objects. They can be further tuned to support in-memory, out-of-core, or streaming execution of the associated algorithm. Thus, analytics applications are characterized by diverse requirements but share a common focus on the application of advanced mathematical modeling, typically on large data sets.

Given the diverse and demanding requirements of analytics and the new technology available in systems, it is imperative to perform an in-depth study of various analytics applications. Any insights would help us identify: (1) optimization opportunities for analytics applications on existing systems and (2) features for future systems that match the requirements of analytics applications. Toward achieving these goals, as the first step, we examine the functional workflow of analytics applications from the usage to the implementation stages. To motivate the study of analytics workloads, we first describe in detail a recent noteworthy analytics application: the Watson intelligent question/answer (Q/A) system.

1.3.1   THE WATSON DEEPQA SYSTEM

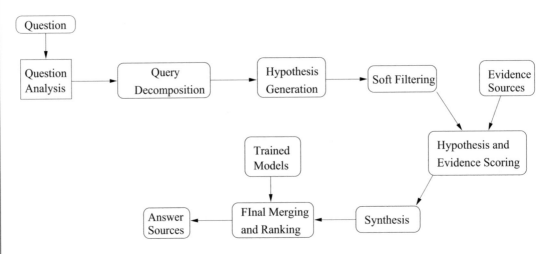

Figure 1.1: Functional workflow of the Watson question-answer system.

Watson is a computer system developed to play the Jeopardy! game show against human participants. Its goals are to correctly interpret the input natural language questions, accurately predict answers to the input questions and finally, intelligently choose the input topics and the wager amounts to maximize the gains. Watson is designed as an open-domain Q/A system using the DeepQA system, a probabilistic evidence-based software architecture whose core computational principle is to assume and pursue multiple interpretations of the input question, to generate many plausible answers or hypotheses and to collect and evaluate many different competing evidence paths that might support or refute those hypotheses through a broad search of large volumes of content. This process is accomplished using multiple stages. The first stage, question analysis and decomposition, parses the input question and analyzes it to detect any semantic entities like names or dates. The analysis also identifies any relations in the question using pattern-based or statistical approaches. Next, using this information, a keyword-based primary search is performed

over a varied set of sources, such as natural language documents, relational databases and knowledge bases, and a set of supporting passages (initial evidence) is identified. This is followed by
the candidate (hypothesis) generation phase which uses rule-based heuristics to select a set of
candidates that are likely to be the answers to the input question. The next step, hypothesis and
evidence scoring, for each evidence-hypothesis pair, applies different algorithms that dissect and
analyze the evidence along different dimensions of evidence such as time, geography, popularity,
passage support, and source reliability. The end result of this stage is a ranked list of candidate answers, each with a confidence score indicating the degree to which the answer is believed correct,
along with links back to the evidence. Finally, these evidence features are combined and weighted
by a logistic regression to produce the final confidence score that determines the successful candidate (i.e., the correct answer). In addition to finding correct answers, Watson needs to master
the strategies to select the clues to it's advantage and bet the appropriate amount for any given
situation. The DeepQA system models different scenarios of the Jeopardy! game using different
simulation approaches (e.g., Monte Carlo techniques) and uses the acquired insights to maximize
Watson's winning chances by guiding topic selection, answering decisions, and wager selections.

1.3.2   FUNCTIONAL FLOW OF ANALYTICS APPLICATIONS

The Watson system displays many traits that are common across analytics applications. They all
have one or more functional goals. These goals are accomplished by one or more multi-stage
processes, where each stage is an independent analytical component. As Figure 1.2 illustrates,
execution of an analytics application can be partitioned into three main components: (1) solution,
(2) library, and (3) implementation.

The solution component is end-user focused and uses the library and implementation components to satisfy user's functional goal, which can be one of the following: prediction, prescrip-
tion, reporting, recommendation, quantitative analysis, pattern matching, or alerting (Davenport
and Harris [2007], Davenport et al. [2010]). For example, Watson's key functional goals are: pattern matching for input question analysis, prediction for choosing answers, and simulation for
wager and clue selection. Usually, any functional goal needs to be achieved under certain runtime
constraints, e.g., calculations to be completed within a fixed time period, processing very large
datasets or large volumes of data over streams, supporting batch or ad-hoc queries, or supporting
a large number of concurrent users. For example, for a given clue, the Watson system is expected
to find an answer before any of the human participants in the quiz. To achieve the functional and
runtime goals of an application, the analytical solution leverages well-known analytical disciplines
such as machine learning, data mining, optimization, data analysis, and simulation and modeling. As we will observe, there is also substantial overlap between these disciplines. For example,
statistics, data mining, and machine learning disciplines are closely related. Both data mining
and machine learning use statistical techniques; data mining extracts information via discovering
known patterns from existing data sources (e.g., finding patterns in customer sales data), whereas
machine learning *learns* by building a model of the underlying system and use it either to answer

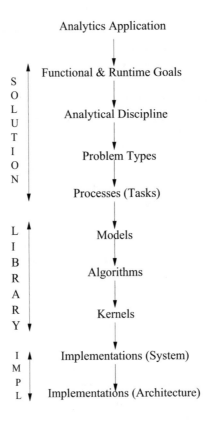

Figure 1.2: Simplified functional workflow of analytics applications.

an unknown input query (e.g., recognizing handwritten characters) or find hidden relationships. Each analytical discipline can support different problems types: e.g., statistics covers descriptive and inferential statistics, data analysis includes both structured and unstructured analysis, machine learning and data mining approaches include unsupervised or supervised learning in which the learning process uses a training dataset with a set of records, where each record consists of values of key input factors that impact the model, called *features*, and a *label* which represents the corresponding result.

In practice, an analytics solution is built as a process with four distinct phases: data ingestion and pre-processing of input data, extract and transform input data to select appropriate data, build an analytical model using the selected data, and then use this model to compute the final decision. Each phase can have one or more tasks which are implemented using appropriate models from the analytics disciplines. These tasks are linked to build the end-to-end solution. Table 1.3 lists the analytics disciplines that are used to achieve functions goals of key business focus areas. As illustrated in Table 1.3, in many cases, a functional goal can be achieved by using more than one problem types. The choice of the problem type to be used depends on many factors that include runtime constraints, underlying software and hardware infrastructure, etc. For example, customer churn analysis is a technique for predicting the customers that are most likely to leave the current service provider (retail, telecom, or financial) for a competitor. This technique can make use of one of the three analytics disciplines: statistics, machine learning, or data analysis. One approach models individual customer's behavior using various parameters such as duration of service, user transaction history, etc. These parameters are then fed either to a statistical model such as regression or to a machine learning model such as a decision tree, to predict if a customer is likely to defect (Mutanen [2006]).

Table 1.3: Examples of analytical focus areas, functional goals, and corresponding analytical disciplines

Focus Areas	Functional Goals	Analytical Problem Type
Revenue Prediction	Prediction	Supervised Learning
BioSimulation	Prediction	Rule Engines and Simulation
Product Portfolio Optimizations	Prescription	Combinatorial Optimizations
Financial Performance Prediction	Prediction	Data Mining
Disease Spread Prediction	Prediction	Supervised Learning
Topic/Semantic Analysis	Pattern Matching	Text Analytics
Semiconductor Yield Analysis	Prediction	Data Mining
Cross-Sell Analysis	Recommendation	Data Mining
Anomaly Detection	Alerting	Data Mining and Data Analysis
Risk Analysis	Quantitative Analysis	Supervised Learning
Retail Sales Analysis	Reporting	Supervised Learning
BioInformatics	Quantitative Analysis	Data Analysis

The second approach models behavior of a customer based on her interactions with other customers. This strategy is commonly used in the telecom sector, where customer calling patterns are used to model subscriber relationships as a graph. This unstructured graph can then be analyzed to identify subscriber groups and their influential leaders: usually the active and well-connected subscribers. These leaders can then be targeted for marketing campaigns to reduce defection in the members of her group (Nanavati et al. [2006]).

The library component is usually designed to be portable and broadly applicable (e.g., the DeepQA runtime that powers the Watson system). A library usually provides multiple implementations of specific models of the problem types shown in Table 1.3. For a problem type, a solution is built using one or more processes or tasks. For example, an unsupervised learning problem can be solved using one of the two tasks: classification or clustering (Han and Kamber [2006]). Each task can be then implemented by different analytical models, each of which can in turn use one or algorithms. For instance, the associative mining model can be implemented using the association rule mining algorithms or using decision trees (Chapter 2). Similarly, classification can be implemented using nearest-neighbor, neural network, or naive Bayes algorithms (Chapter 2). It should be noted that, in practice, the separation between models and algorithms is not strict and, many times, the same algorithm can be used for supporting more than one models. For instance, neural networks can be used for clustering or classification.

Finally, depending on how the problem is formulated, each algorithm uses different data structures and kernels. For example, many algorithms formulate the problem using dense or sparse matrices and invoke kernels like matrix-matrix, matrix-vector multiplication, matrix factorization, and linear system solvers. These kernels are sometimes intensively optimized for the underlying system architecture, in libraries such as IBM ESSL or Intel MKL. Any kernel implementation can be characterized according to: (1) how it is implemented on a system consisting of one or more processors, and (2) how it optimized for the underlying processor architecture. The system-level implementation varies depending on whether a kernel is sequential or parallel, and if it uses in-memory or out-of-core data. Many parallel kernels can use shared or distributed memory parallelism. In particular, if the algorithm is embarrassingly parallel, requires large data, and the kernel is executing on a distributed system, it can often use the map-reduce approach (Dean and Ghemawat [2010]). At the lowest level, the kernel implementation can often exploit hardware-specific features such as short-vector data parallelism (SIMD) or task parallelism on multi-core CPUs, massive data parallelism on GPUs, and application-specific parallelism on FPGAs or ASICs.

Although analytics applications have come of age, they have not yet received significant attention from the computer architecture community. It is important to understand systems implications of the analytics applications, not only because of their diverse and demanding requirements, but also, because systems architecture is currently undergoing a series of disruptive changes. Widespread use of technologies such as multi-core processors, specialized co-processors or accelerators, flash memory-based solid state drives (SSDs), and high-speed networks has created new optimization opportunities. More advanced technologies such as phase-change memory are on the horizon and could be game-changers in the way data is stored and analyzed. In spite of these trends, currently there is limited usage of such technologies in the analytics domain. Given the wide variety of algorithmic and system alternatives for executing analytics applications, it is often difficult for solution developers to make the right choices to address specific problems. Naive usage of modern technologies often leads to unbalanced solutions that further increase optimization complexity. Thus, to ensure effective utilization of system resources, CPU, memory, networking, and storage, it is necessary to evaluate analytics workloads in a holistic manner.

We aim to understand the application of modern systems technologies to optimizing analytics workloads by exploring the interplay between overall system design, core algorithms, software (e.g., compilers, operating system), and hardware (e.g., networking, storage, and processors). Specifically, we are interested in isolating repeated patterns in analytical applications, algorithms, data structures, and data types, and using them to make informed decisions on systems design. Toward this goal, we have been examining the functional flow of a variety of analytical workloads across multiple domains (Table 1.1). As a result of this exercise, we have identified a set of commonly used analytical models, called *analytics exemplars* (Bordawekar et al. [2011]). We believe that these exemplars represent the essence of analytical workloads and help us identify *com-*

mon computational and runtime patterns across different analytics workloads. Thus, the analytics exemplars can be used as a toolkit for performing exploratory systems design for the analytics domain. Table 2.1 lists the 14 analytics exemplars and the corresponding functional domains. In this book, we rely on these exemplars to illustrate that analytics applications benefit greatly from holistically co-designed software and hardware solutions and demonstrate this approach using examples from different domains.

1.4 INTENDED AUDIENCE

This book is designed for those with a background in computer architecture and compiler design. The goal of this book is to provide a high-level survey of key analytics models and algorithms, without going into mathematical details. The rest of this book is organized as follows: we first overview the 14 analytics exemplars and the key algorithms used by each exemplar; we then summarize computational and runtime patterns of these exemplars. Based on this information, we discuss various acceleration opportunities and demonstrate them using case studies of key exemplars. We conclude by discussing various open issues in accelerating analytics workloads, e.g., new architectural features for supporting analytics workloads, impact on programming models and runtime systems, and designing analytics systems. We hope this study acts as a *call to action* for computer architects and systems designers to focus future research on analytics.

Further Reading

Using analytics in business problems, see Anderson et al. [2009], Apte et al. [2002], Business Week [2006], Cantor et al. [1997], Crosbie and Bohn [2003], Davenport and Harris [2007], Davenport et al. [2010], Eckerson [2009], Goode [2011], Harding et al. [2006], Madsen [2009], Manyika et al. [2011], NIPS 2009 Workshop [2009], Nisbet et al. [2009], Piatetsky-Shapiro [2011], Rexer [2013], Science Special Issue [2011], Shmueli et al. [2010], Suman [2006], The Economist [2007]; **Analytics solutions: Netflix** Bell et al. [2010], **Pandora** Glaser et al. [2002], Joyce [2006], **Watson** Ferrucci et al. [2010], **Telecom Churn Prediction** Dasgupta et al. [2008], Ngai et al. [2009], Richter et al. [2010], **Cognos Consumer Insight (CCI)** IBM Institute for Business Value [2011], Sindhwani et al. [2011], **UPS** Armacost et al. [2004], Lohatepanont and Barnhart [2004], **Hyperic and Splunk** Splunk Inc. [2011]; **Analytics products: R, Weka** Hall et al. [2009], **STATISTICA, RapidMiner, Oracle, SAP, and IBM** Bhattacharya et al. [2009], IBM Corp. [2011]; **Academic references** Bekkerman et al. [2011], Han and Kamber [2006], Leskovec et al. [2014], StatSoft Inc. [2010], Wu et al. [2008].

CHAPTER 2

Overview of Analytics Exemplars

2.1 EXEMPLAR MODELS

Table 2.1 presents the 14 analytics exemplars with their application domains and Table 2.2 lists the associated analytics problem types. As these tables illustrate, each analytics exemplar can be used in multiple domains and can address multiple functional goals. Further, an application domain can use one or more exemplars (e.g., marketing). These analytic models span the key analytics disciplines such as data mining, machine learning, statistics, simulation, and data analysis. Finally, an exemplar can be used in one or more analytics phases, e.g., the data ingestion and preprocessing phase can use text analytics or time-series processing, the transform and load phase can use on-line analytical processing or graph analytics, the model building phase can use regression, clustering, decision tree, and finally the decision processing can use mathematical programming, Monte Carlo methods, decision trees, or graph analytics. In the remainder of this chapter we discuss these models in detail; for every model, we first discuss the target application domains, then outline the basic idea, and then summarize the key implementation algorithms.

Table 2.1: Analytics exemplar models and their application domains

Analytics Exemplar	Key Application Domains
Regression Analysis	Social Sciences, Marketing, Economics
Clustering	Marketing, Medical Imaging, Document Management
Nearest-Neighbor Search	Computational Biology, Image Processing
Associative Rule Mining	Retail Analysis, Bio-Informatics, Intrusion Detection
Recommendation Systems	Online Media and e-commerce
Neural Networks	Image and Speech Recognition, Fraud Detection
Support Vector Machines	Bio-Informatics, Document Classification, Financial Modeling
Decision Tree Learning	Medical Diagnostics, Fraud Detection, Marketing, Manufacturing
Time Series Processing	Medical Informatics, Geology, Economics
Text Analytics	Web Search, Medical Informatics, Bio-Informatics
Monte Carlo Methods	Computational Finance, Insurance Risk Modeling
Mathematical Programming	Routing, Scheduling, Manufacturing
On-line Analytical Processing	Sales/Marketing, Retail,
Graph Analytics	Social Analysis, Computational Neuroscience, Logistics

Table 2.2: Analytics exemplars and corresponding problem types

Analytics Exemplar	Analytics Problem Type
Regression	Inferential Statistics, Supervised Learning
Clustering	Data Mining, Unsupervised Learning
Nearest Neighbor Search	Data Mining, Unsupervised Learning
Association Rule Mining	Unsupervised Learning
Recommender Systems	Unsupervised Learning
Neural Networks	Machine Learning, Supervised/Unsupervised Learning
Support Vector Machines	Machine Learning, Supervised Learning
Decision Trees	Machine Learning, Supervised Learning
Text Analytics	Data Analysis, Supervised/Unsupervised Learning
Time Series Processing	Data Analysis, Unsupervised Learning
Mathematical Programming	Mathematical Optimization
Monte Carlo Methods	Simulation
Online Analytical Processing	Data Analysis
Graph Analytics	Data Analysis

2.2 REGRESSION ANALYSIS

Regression analysis is a classical statistical technique used to model the relationship between a dependent variable and one or more independent variables. Specifically, regression can predict how the value of the dependent variable can change when any one of the independent variables is varied while the remaining independent variables are fixed. Regression analysis is primarily used for prediction and forecasting purposes and also to discover relationships between dependent and independent variables, e.g., to estimate conditional expectation of the dependent variable given multiple independent variables. Thus, regression can be viewed as an example of a classifier which uses supervised learning which uses independent variables for training. Regression analysis has been used in many application domains, including economics, psychology, social sciences, marketing, healthcare, and computational finance, e.g., for predicting house prices based on information such as crime rate, population, number of rooms, property tax; for predicting airfares on new routes using the location and airport information such as population, average income, passenger estimates, number of airport gates, etc. (Shmueli et al. [2010]).

Basic Idea: Formally, any regression model relates a dependent variable Y to a regression function f of independent variables (*regressors*) X and unknown parameters, $\boldsymbol{\beta}$: $Y \approx f(X, \boldsymbol{\beta})$. The quality of the predication depends on the amount of information available about the independent variable X. If k is the length of the vector of unknown parameters β, then the regression analysis is possible only if $N \geq k$, where N is the number of observed data points of the form (Y, X). If $N = k$, and the regression function f is linear, then the equations $Y = f(X, \boldsymbol{\beta})$ can be solved exactly. However, if the regression function is nonlinear, then either many solutions exist or a solution may not exist. When $N > k$ (also called an over-determined system), a best-fit strategy is usually used to predict the values of X. We now discuss some of the key regression algorithms.

Linear Regression: Models the relationship between a scalar dependent variable Y and one or more regressor variables X using linear regression functions. A dependent variable y_i can be approximated as a linear combination of regressors x_i. The following represents a simple linear regression model for n data points with *independent* variables x_{ij}, and *regression* coefficients β_i:

$$y_i = \beta_1 x_{il} + \cdots + \beta_p x_{ip} + \epsilon_i = \vec{x}_i^T \beta + \epsilon_i \quad i = 1, \ldots, n, \tag{2.1}$$

where T denotes the transpose, so that $x_i^T \beta$ is the inner product between vectors x_i and β. Any solution for the linear regression model aims to estimate and infer the values of the regression coefficients β_i using estimation methods such as Ordinary or Generalized Least Squares methods.

Nonlinear Regression: The nonlinear regression model is characterized by the fact that the dependent variables Y are related to the regressor variables X via a nonlinear relationship on one or more unknown parameters, β. A nonlinear regression model has the following form:

$$y_i = f(x_i, \beta) + \epsilon_i, \quad i = 1, \ldots, n, \tag{2.2}$$

where the function $f(x_i, \beta)$ is nonlinear as it cannot be expressed as a linear combination of the parameters, β and ϵ_i are random errors. Common nonlinear functions include exponential decay/growth, logarithmic, trigonometric, power, and Gaussian functions.

For nonlinear regression, regression parameters β can be estimated by minimizing a suitable goodness-of-fit expression with respect to β. One popular approach is to minimize the sum of squared residuals using the nonlinear least squares method that uses the Gauss-Newton numerical method. In some cases, maximum likelihood or weighted least squares estimation is used. Alternatively, in some cases, the nonlinear function can be transformed to a linear model. The transformed model can then be estimated using linear regression approaches.

Logistic Regression: The logistic regression is used for prediction of the *probability* of occurrences of an event by fitting data to a logistic function, $f(z) = \frac{1}{1+e^{-z}}$. The variable z is a measure of the total contributions of all the independent variables while $f(z)$ represents probability of a particular outcome, given the set of independent variables. The variable z is usually defined as

$$z = \beta_0 + \beta_1 x_1 + \cdots + \beta_k x_k \tag{2.3}$$

and for any value of z, the output $f(z)$ varies between 0 and 1.

The parameters β_i can then be estimated by maximum likelihood approach via the iteratively re-weighted least squares method that can use the Gauss-Newton numerical algorithm.

Logistic regression is one of the most widely used analytics algorithms both as a standalone method for binary or multi-nominal classification or as a kernel in other classifiers such as neural networks (Bengio et al. [2014], Rexer [2013]).

Further Reading
Algorithms: Han and Kamber [2006], Wu et al. [2008]; **Packages:** R (The R Foundation), Weka (Hall et al. [2009]), IBM SPSS (IBM SPSS [2010b], RapidMiner Rapid-i), STATISTICA (The StatSoft Inc.), Oracle Data Miner (Oracle Corp. [b]), and SAS (SAS Institute Inc.); **Applications:** Shmueli et al. [2010], Smyth [2002], StatSoft Inc. [2010].

2.3   CLUSTERING

Clustering is a process of grouping together entities from an ensemble into classes of entities that are similar in some sense. Clustering is also called *data segmentation* (Han and Kamber [2006], Shmueli et al. [2010]) as it partitions large datasets into segments of *similar* and *dissimilar* datasets. Clustering is an example of unsupervised machine learning and is being used in a wide variety of applications, e.g., in market segmentation for partitioning the customers according to gender, interests, etc., in gene sequence analysis to identify gene families, in medical imaging for differentiating key features in PET scan images, and in clustering documents based on semantic information.

Basic Idea: Any clustering algorithm needs to effectively identify and exploit relevant similarities in underlying potentially disparate data sources. The similarity can be expressed using either geometric *distance*-based metric (e.g., using either Euclidean or Minkowski metric; Kruskal [1964]) or conceptual relationships in the data. The input data can be noisy, of different types (intervalbased, binary, categorical, ordinal, vector, mixed, etc.), have high dimensions, and have large size (e.g., millions of objects). Thus, the key challenges before any clustering algorithm are the effective use of the similarity metric, exploitation of intrinsic characteristics of the data, support for large number of dimensions, large data sets, and different cluster shapes.

Current approaches to solving the clustering problem can be broadly classified into parametric and nonparametric approaches. The parametric or model-based methods assume that the input data is associated with a certain probability distribution and the clustering to designed to fit the data to some mathematical model (Han and Kamber [2006]). The nonparametric methods exploit spatial properties (e.g., distance or density) of the input data. Clustering algorithms also differ depending on the dimensionality of the datasets (Kriegel et al. [2009]). We now discuss some of the key clustering algorithms; K-Means and Hierarchical clustering algorithms are examples of nonparametric clustering, and EM clustering is an example of parametric clustering.

K-Means Clustering: The *k-means* clustering algorithm (Hartigan and Wong [1979], MacQueen [1967]) is an example of a partitioning method that constructs k partitions from a database of n objects, where each partition represents a cluster and $k \leq n$. Each cluster contains at least one object and an object lies in only one cluster. A partitioning algorithm creates an initial assignment of objects to partitions and then iterative relocation technique are used (Han and Kamber [2006]) to move objects among the groups. The objects in a group are considered to be closer to each than objects in different clusters. In the k-means algorithm, the cluster similarity is measured as the mean value of objects, which can be viewed as the cluster's *centroid* or center of gravity.

The k-means algorithms works only on data whose mean can be computed (e.g., it is not possible to compute a mean value for categorical datasets). The algorithm also requires the number of clusters k to be defined a priori. The algorithm cannot discover clusters with different shapes and is sensitive to noise and outliners as they can significantly affect calculations of the mean value. One variation of the k-means problem, the **k-modes** method uses modes as a measure of similarity for categorical objects and a frequency-based update method (Chaturvedi et al. [2001]).

Hierarchical Clustering: Hierarchical clustering methods group data objects into a tree of clusters. Hierarchical clustering often produces data clusters that can be viewed graphically using a *dendrogram*. Hierarchical methods can be classified into: (1) agglomerative, i.e., those use a bottom-up strategy to construct increasing large clusters until certain termination conditions are met; and (2) divisive, i.e., those that start from a single cluster and subdivide it into smaller pieces until termination conditions are met.

The most popular hierarchical clustering algorithm, BIRCH (Zhang et al. [1996]), uses the agglomerative strategy. BIRCH uses a two-step strategy to improve I/O scalability and clustering flexibility. In the first *micro-clustering* stage, it uses the hierarchical clustering strategy to build an initial set of in-memory clusters, that summarize the information in the original data. The second *macro-clustering* phase processes these summarized in-memory clusters (i.e., it does not fetch the raw data again) using any clustering method, e.g., iterative partitioning, and computes the final clustering.

EM Clustering: The Expectation-Maximization (EM) clustering is a model-based (parametric) approach that extends the k-means partitioning algorithm. The EM clustering algorithm assumes that the underlying data is a mixture of the k probability distributions (referred to as *component* distributions), where each distribution represents a cluster. The key problem for any model-based algorithm is to estimate the parameters of the probability distributions so as to best fit the data.

The EM algorithm (Dempster et al. [1977], Han and Kamber [2006]) is an iterative refinement algorithm that extends the k-means paradigm (the EM algorithm also uses a predetermined number of clusters): it assigns a data item to a cluster according to a weight representing the probability of membership (unlike the cluster mean metric used in the k-means algorithm). The new means of these clusters are then computed using the weighted measures. The most common version of the EM algorithm learns a mixture of Gaussian distribution (Bilmes [1998]). It starts with an initial estimate of the parameters of the mixture model, referred to as the parameter vector. Each data item is assigned a probability that it would possess a certain set of attributes given that it was a member of a given cluster. The items are then re-scored against the mixture density produced by the parameter vector and then the items are used to update the parameter estimates. The complexity of the EM algorithm is linear in d (the dimensions or features of the input data), n (the number of data items), and t (the number of iterations) ($O(dnt)$).

Further Reading

Algorithms: K-means (Chaturvedi et al. [2001], Hartigan and Wong [1979], MacQueen [1967]), Hierarchical Clustering (Chiu et al. [2001], Zhang et al. [1996]), EM (Dempster et al. [1977]), PROCLUS (Aggarwal et al. [1999]) and CLIQUE (Agrawal et al. [1998]); **Packages:** R (The R Foundation), Weka (Hall et al. [2009]), IBM SPSS and InfoSphere DataMining (IBM Corp. [c], IBM SPSS [2010a]), RapidMiner (Rapid-i), STATISTICA (The StatSoft Inc.), Oracle Data Miner (Oracle Corp. [b]) and SAS (SAS Institute Inc.); **Applications:** Han and Kamber [2006], Nisbet et al. [2009], Shmueli et al. [2010].

2.4   NEAREST NEIGHBOR SEARCH

Nearest neighbor search is an optimization problem for finding the closest points in a metric space. This problem of identifying points from an ensemble of points that are in some defined proximity to a given query point has been applied for classification and clustering purposes in multiple application domains such as distributed systems, image processing, data mining, computational biology, data compression, and machine learning. The notion of proximity varies from domain to domain and is usually formulated using a suitable metric function (e.g., Euclidean distance for spatial proximity). For example, the online media providers such as Netflix and Pandora use nearest neighbor algorithms to suggest movies or songs that match a particular taste of a particular user (Bell et al. [2010], Joyce [2006]). Nearest neighbor algorithms have also been used for finding similarities in multi-media data (e.g., related images or videos) to detect any copyright violations (Boiman et al. [2008], Indyk [2004]). Other well-known applications of the nearest neighbor search algorithm include control systems, robotics, and drug discovery (Beyer et al. [1999], Jadbabaie et al. [2003], Stanton et al. [1999]).

Basic Idea: Formally, the nearest neighbor problem can be defined as follows. Given a set S of n points in some metric space (X, d), the problem is to preprocess S so that given a query point $p \in X$, one can efficiently find a $q \in S$ that minimizes $d(p, q)$. In practice, several variations of this definition are implemented as per input data characteristics and runtime constraints. Broadly, the existing sequential nearest neighbor solutions can be classified as per the dimensionality of input data, type of metric used in proximity calculations, result cardinality (e.g., top-k and all-pairs), and data size (e.g., Terabyte datasets).

The key aspect of any nearest neighbor algorithm is the metric function used for calculating the proximity distance between the input data points. The most widely used metric in the nearest neighbor algorithms is the Euclidean distance: distance between any two points p_i and p_j in a d-dimensional space can be computed as $|\vec{p_i} - \vec{p_j}| = \sqrt{\sum_{k=1}^{k=d} (p_{ik} - p_{jk})^2}$, where p_{ik} is the kth component of the vector $\vec{p_i}$. In practice, the Euclidean distance has proven to be effective for low-dimensional data. For high-dimensional data, more generalized forms of metric distances are employed (e.g., Hamming distance; Uhlmann [1991]). In the case of two-dimensional data, the nearest neighbor problem can be solved by using Voronoi diagrams (Omohundro [1987]). As the dimensionality become very high, the distance calculations become ineffective by the *curse of dimensionality*. For datasets with very high dimensionality (e.g., in computer vision), these search algorithms provide sub-linear performance. One approach to deal with this inefficiency is to define an *approximate* version of the nearest neighbor problem: this version of the problem identifies points from an ensemble of points whose distance from the given query point is no more than $(1 + \epsilon)$ times the distance of the true k^{th} nearest-neighbor.

K-d Trees: The k-d tree algorithm addresses the precise nearest neighbor problem (Bentley [1975, 1980], Friedman et al. [1977]). The k-d tree is a binary tree used for representing k-dimensional data using recursive hyperplane decomposition. Each node of the k-d tree represents a region of the input dataset and its partitioning. Each level of the k-d tree covers the entire dataset. In k dimensions, a record is represented by k keys, where each key represents a position in the kth dimension. The k-d tree is then constructed by recursively selecting one of the k coordinates as the discriminator dimension and then partitioning the dataset into the subset of vectors according to a certain partition value. During the query process, the tree is recursively traversed from the root: at every level, the value of the discriminator coordinate of the query record is compared against the partition value and either the left or right path is chosen for further traversal. When the traversal reaches a leaf node, the query record is compared with the records in the leaf, and a list of the m closest records is maintained. Overall, for a dataset of N k-dimensional records, the tree search requires $O(logN)$ time, with $O(N)$ space consumption. The k-d tree is very effective for small dimensionalities. As the number of dimensions increases, the quality of discrimination degrades as the proximity calculations are based only on a subset of coordinates.

Approximate Nearest Neighbor (ANN): The ANN algorithm uses hierarchical space decomposition for solving the approximate nearest neighbor problem for low-dimensional data. The ANN algorithm represent points in a d-dimensional space using a balanced box-decomposition (BBD) tree with O(log n) depth (Arya et al. [1998]). ANN recursively sub-divides the space into a collection of *cells*, each of which is either an axis-aligned d-dimensional *fat* (i.e., the ratio between the longest and shortest sides is bounded) rectangle or the set-theoretic difference of two rectangles, each enclosed within the other. Each node of the tree is associated with a cell. Thus, it is associated with all points contained within the enclosed cells. Each leaf cell is associated with a single point lying within the bounding rectangle of the cell. The leaves of the tree span the entire space. The ANN tree has $O(n)$ nodes and can be built in $O(dn \log n)$ time. During the querying process, for a given query point q, a priority queue of the internal nodes of the BBD-tree is created, where priority of a node is inversely related to the distance between the query point and the cell corresponding to the node. The highest priority node is then selected for recursive descent toward the leaves. As the descent progresses, the priority queue is updated appropriately. Let p denote the closest point seen so far; as soon as the distance from q to the current leaf exceeds $\frac{disi(q,p)}{(1+\epsilon)}$, the search can be terminated.

Locality Sensitive Hashing (LSH): The locality-sensitive hashing (LSH) algorithms are designed for solving the approximate nearest neighbor problem for very high-dimensional data sets (e.g., with a million feature vectors). The key idea behind the LSH algorithms is to hash points using several hash functions to ensure that for each function, the probability of collision is much higher for points that are close to each other than for those that are far apart (Andoni and Indyk [2008], Indyk and Motwani [1998]). Then for the query point, one can determine its nearest neighbors by the hashing the query point and retrieving the points in the bucket containing that point. The LSH method relies on a family of hash functions that have the property that if two points are *close*, then they hash to same bucket with high probability; if they are *far apart*, they hash to the same bucket with low probability.

Further Reading
Basic idea: Omohundro [1987], Uhlmann [1991]; **Algorithms**: KD-Tree (Bentley [1975, 1980], Friedman et al. [1977]), ANN (Arya et al. [1998]); LSH (Andoni and Indyk [2008], Datar et al. [2004], Indyk and Motwani [1998], Wang et al. [2014]), Ball Trees (Omohundro [1987, 1989, 1991]), Metric Trees (Ciaccia et al. [1997], Liu et al. [2006], Moore [2000]), Spill Trees (Liu et al. [2004, 2007]), and Cover Trees (Beygelzimer et al. [2006]); **Packages**: ANN (Mount and Arya [2010]); **Applications**: Bell et al. [2010], Beyer et al. [1999], Han and Kamber [2006], Jadbabaie et al. [2003], Joyce [2006], Nisbet et al. [2009], Stanton et al. [1999].

2.5   ASSOCIATION RULE MINING

Association rule mining is a key data mining method used for discovering relationships between variables. Agrawal et al. [1993] first proposed using association rule mining for identifying relationships between items purchased in retail stores, a process widely known as the market-basket analysis. Over the years, this method has been applied to more complex data patterns such as sequences, trees, graphs, etc., and different application domains such as bio-informatics, intrusion detection, and web-usage analysis (Agrawal and Srikant [1994], Agrawal et al. [1993], wiki, Wu et al. [2008]).

Basic Idea: Formally, the association rule mining processes a set (or database) D of transactions, where each transaction T is a set of items such that $T \subseteq I$, where $I = \{i_1, i_2, \cdots, i_m\}$ is a set of literals, called items. Each transaction is associated with a unique identifier, called

TID. By *association* rule, we mean an implication of the form $X \Longrightarrow I_j$, where X is a set of some items in I and I_j is a single item in I that is not present in X. The rule $X \Longrightarrow Y$ in the transaction set D has a *confidence c* if $c\%$ of transactions in D that contain X also contain Y. The rule $X \Longrightarrow Y$ has *support s* in the transaction set D if $s\%$ of transactions in D contain $X \cup Y$ (Agrawal and Srikant [1994], Agrawal et al. [1993]). While confidence is a measure of the rule's strength, support corresponds to statistical significance. The support of a rule $X \Longrightarrow Y$ is defined as $supp(X \Longrightarrow Y) = supp(X \cup Y)$. The confidence of this rule is defined as $conf(X \Longrightarrow Y) = supp(X \cup Y)/supp(X)$. Given a set of transactions D, the problem of mining association rules is to generate all association rules that have support and confidence greater than the user-specified minimum support (called *minsup*) and minimum confidence (called *minconf*). The problem of discovering all association rules can be decomposed into two subproblems (Agrawal and Srikant [1994], Agrawal et al. [1993], Hipp et al. [2000]).

- Find all sets of items (*itemsets*) that have transaction support above *minsup*. The support for an itemset is the number of transactions that contain the itemset. Itemsets with at least *minsup* are called large (frequent) itemsets, and all others are small itemsets.

- Use the frequent itemsets to discover the desired rules. For every frequent itemset l, we find all non-empty subsets of l. Note that the support attribute follows the downward closure property: all subsets of a frequent itemset must be frequent. For every such subset a, we output a rule of the form $a \Longrightarrow (l - a)$ if the ratio of support(l) to support(a) is at least *minconf*. One needs to consider all subsets of l to generate rules with multiple consequences. The number of rules can grow exponentially with the number of items, but the choices can be pruned using both *minsup* and *minconf*.

Once the associated rules are computed, further pruning may be required to select the most useful rules (Klemettinen et al. [1994]).

The key task in the association rule mining process is to find all itemsets that are frequent with respect to a given minimal threshold *minsupp*. All existing associative rule mining algorithms employ the downward closure property of the itemset support for pruning the search space: every subset of a new frequent itemset must be frequent. These algorithms can be classified by how the search space is traversed to construct itemsets (Hipp et al. [2000]): the breadth-first search (BFS) algorithms compute all itemsets of size $k - 1$ before building the itemsets of size k, while the depth-first search (DFS) algorithms, hierarchically compute all possible itemsets of size k from a list of frequent itemsets of size j ($j < k$), before processing other itemsets of size j.

Frequent and potentially frequent itemsets are called *candidate* itemsets. There are two common ways of computing support values of these candidate itemsets. The first approach directly counts occurrences of that itemset in all D transactions. The second approach uses *set intersection* to compute the support values of the itemsets. For every item in the transaction set, a list of identifiers (TIDs) that correspond to the transactions containing that item is maintained (*tidlist*). Accordingly, tidlists also exist for every itemset X and denoted by $X.tidlist$. The tidlist of a candidate $C = X \cup Y$ can be obtained as $C.tidlist = X.tidlist \cap Y.tidlist$. The actual support of the itemset C can then be computed as $|C.tidlist|$.

Based on the strategy for traversing the candidate search space (i.e., DFS vs. BFS) and for computing the support values (i.e., direct counting vs. intersection), the association rule mining algorithms can be partitioned into four key families (Hipp et al. [2000]) (Table 2.3). Both *Apriori* and *Partition* algorithms exploit the downward closure of itemsets by iteratively computing large

Table 2.3: Classification of associated rule mining algorithms

Traversal Method	Support Computations	Algorithm
Breadth-first search	Direct Counting	Apriori
Breadth-first search	Intersection	Partition
Depth-first search	Direct Counting	FP-Growth
Depth-first search	Intersection	Eclat and MaxClique

candidate itemsets using smaller sized candidate itemsets (Agrawal and Srikant [1994], Savasere et al. [1995]). While the Apriori algorithm makes multiple passes over the raw data, the Partition algorithm requires only two passes as it partitions the data into non-overlapping partitions such that the number of itemsets to be evaluated can be fit into the main memory. The FP-growth algorithm use depth-first traversal to build an extended prefix-tree structure, called a FP-tree, that compactly represents groups of frequent items as paths (Han et al. [2000]). The item-set mining algorithm then traverses FP-tree to compute frequent patterns. The Eclat and MaxClique family of algorithms view the creation of candidate itemsets as a lattice that represents structural relationships among candidate itemsets, and use two strategies, one based on equivalence classes and another on maximal cliques in a hypergraph to predict candidate itemsets. These itemsets logically induce a sub-lattice that is then traversed depth-first to generate all frequent itemsets (Zaki et al. [1997], Zaki [2000]).

Further Reading

Basic idea: Agrawal and Srikant [1994], Agrawal et al. [1993], Hipp et al. [2000], Klemettinen et al. [1994]; **Algorithms:** Apriori (Agrawal and Srikant [1994]), Partition (Savasere et al. [1995]), FP-Growth (Han et al. [2000, 2004]), Eclat and MaxClique (Zaki et al. [1997], Zaki [2000]); **Packages:** R (The R Foundation), Weka (Hall et al. [2009]), RapidMiner (Rapid-i), STATISTICA (The StatSoft Inc., SAS SAS Institute Inc.), Microsoft SQL Server (Microsoft Corp.), IBM InfoSphere DataMining (IBM Corp. [c]), Oracle Data Miner (Oracle Corp. [b], and SAP Legler et al. [2006]); **Applications:** Davenport and Harris [2007], Davenport et al. [2010], Han and Kamber [2006].

2.6   RECOMMENDER SYSTEMS

The goal of a recommender system is to predict interest of users for a set of items and based on the interest, provide meaningful recommendations. This method is widely used by a variety of online e-commerce and media companies, e.g., book recommendations on Amazon, movie recommendations by Netflix, or song recommendations from Spotify. Recommender systems differ from associated rule mining in that recommendation systems make their selections based on the *external* information (e.g., collective ratings for a book from multiple readers or user- and item-specific attribute knowledge), rather than using implication rules.

Basic Idea: A recommendation task can be defined as a process to predict likely preferences of an user and make recommendations that match these preferences. Broadly, recommendations can be classified into the following categories: (1) non-personalized recommendations recommend items based on the average rating given by other customers on the items; (2) attribute-based recommendation system recommend products based on the product properties (e.g., genre of a book); (3) item-to-item correlation that recommends items based on the set of items that the customer has already interested in; and (4) people-to-people correlation that recommends items based on the correlation between the customer and other customers that have purchased items (Schafer et al. [1999]). In general, these recommendation system methods can be classified into Collaborative Filtering (CF), Content-based recommendation, and hybrid approaches that combine both methods (Melville and Sidhwani [2010]).

Collaborative Filtering (CF): This approach uses social collaboration to aggregate ratings for items in a domain, and exploits similarities in the ratings for determining if an item can be recommended. The CF methods can be further classified into *neighborhood-based* (or *memory-based*) and *model-based* approaches.

In the neighborhood-based CF approach, a subset of related users (*a neighborhood*) is selected based on their similarity to the given user, and their collective ratings are used to predict for the user. The most common measure of similarity is the Pearson correlation coefficient between two ratings. One can also treat the user ratings as a vector in an m-dimensional space and compute similarity based on the cosine of the angle between them. This approach can be also applied to the item-to-item correlation where rather than matching similar users, a user's rated items are matched to similar items.

The model-based CF approach formulates the CF problem as a statistical model of user ratings and solves it by estimating its parameters. Current model-based algorithms use matrix factorization models to detect similarities between users and items that is induced by some hidden (*latent*) lower dimensional structure present in the data. These models take as an input a $n \times m$ ratings matrix r whose each element r_{ij} represents rating of user i for the item j. The ratings matrix is then factored into two matrices, W and H, where $W = [w_1 \ldots, w_n]^T$ is an $n \times k$ matrix, and $H = [h_1, \ldots, h_m]^T]$ is an $k \times m$ matrix, and k is the number of latent dimensions. The factorization learns k-element feature vectors w_u, h_i such that inner products $w_u^T h_i$ approximate the known preference ratings $r_{u,i}$.

An alternative approach applies the non-negative matrix factorization (NNMF; Lee and Seung [1999]) formulation to the recommendation problem. The non-negative constraints on the factor matrix leads to a part-based interpretation, where the ratings of each user can be viewed as an additive sum of basis vectors of ratings in the item space. In practice, the matrix factorization is implemented using the Alternative Least Squares (ALS) algorithm (Hu et al. [2008]). The ALS algorithm is an iterative algorithm which starting a fixed ratings matrix, alternatively computes one of the factor matrices (W or H), while keeping the other factor matrix constant. At every stage, the matrix values are approximated to minimize the quadratic loss function defined as the sum of squared differences (Zhou et al. [2008]).

Content-based Recommendation: Unlike the collaborative filtering approach that only uses collective ratings of individual users, content-based recommendations finalize their selections based on the representations of content that interest the users and representations of contents of the items (e.g., movie genres).

In practice, this problem is commonly addressed using either *similarity* or *classification* approaches on the associated textual content. In the similarity approach, the content related to user's preference is viewed as query, and the unrated documents related to the items are scored with relevance/similarity to this query. In the classification approach, the classification model is trained using the user rating as a label and content attributes as features. For example, for book recommendation, fields such as title, author, or subject are used to train a multinomial classifier, with k classes, where 1 to k is the scale of ratings (numerical ratings can also be used to train a binary classifier). The classification model can be then implemented using a variety of traditional classifiers such as the Naive Bayes classifier, k-nearest neighbor, or decision trees.

Hybrid Approaches: Hybrid approaches combine the collaborative and content-based recommenders to take advantages of both approaches. A simple hybrid approach uses both approaches to generate two separate rankings of recommendations and then merge the results to compute the final list. The basic approach can be extended to weigh individual rankings, e.g., increase the weight of collaborative component as the number of uses accessing an item increases.

One class of hybrid algorithms maintains content-based profiles of users and uses it for finding similar users. A user-profile matrix is computed (as opposed to the user-ratings matrix) and a collaborative filtering approach is directly applied to this matrix to suggest recommendations. Another approach treats the recommendation process as a classification task, in which both collaborative and content information is used together create features for training a classifier system. A related approach, *content-boosted collaborative filtering*, trains a Naive Bayes classifer using documents that describe rated items of each user and uses predictions from this classifier to boost an existing sparse user ratings matrix. This *pseudo rating matrix* is then used to compute a subset of users (neighbors) that have similar interests.

Further Reading
Basic idea: Adomavicius and Tuzhilin [2005], Koren et al. [2009], Su and Khoshgoftaar [2009]; **Algorithms:** Collaborative Filtering (Breese et al. [1998], Goldberg et al. [1992], Hu et al. [2008], Lee and Seung [1999], Linden et al. [2003], Su and Khoshgoftaar [2009], Zhou et al. [2008]); Content-based Recommendation (Melville et al. [2002], Pazzani [1999], Pazzani and Billsus [1997]); Hybrid Approaches (Basu et al. [1998], Claypool et al. [1999], Good et al. [1999]; **Applications:** Bell et al. [2010], Bhasker and Srikumar [2010], Linden et al. [2003].

2.7   SUPPORT VECTOR MACHINES

Support Vector Machines (SVMs) is a family of supervised learning methods, primarily used for classification and regression analysis (Bennett and Campbell [2000], Vapnik [1995, 1998]). While this technique was originally designed for pattern recognition applications, it has found applications in a wide spectrum of fields, e.g., astrophysics (parameter estimation, red-shift detection), bio-informatics (e.g., gene classification), medical imaging (e.g., brain fMRI processing), text analytics (e.g., string-based text classification), time-series prediction (e.g., traffic modeling), and financial modeling (e.g., stock indices behavior prediction) (Burges [1998], Guyon [2006], Hsu et al. [2010], Noble [2006], Pereira et al. [2009], Rao et al. [2011], Sewell).

Basic Idea: An SVM is a class of algorithms that use the *kernel* mapping approaches to map original data to high-dimensional *feature* space and combine statistical learning approaches with optimization techniques for classifying the input dataset. A classification application usually involves operating on two types of data sets: one for training and another for testing. Each instance of the training set contains one *target value* (i.e., class labels) and several attributes (i.e., features). The goal of the SVM is to produce a model based on the training data that predicts the target values of the tests data, given only the test data attributes. The standard form of SVMs usually classify the tests data into two categories. Intuitively, each data point in the training set is first mapped to a high-dimensional space, which is then partitioned by a set of hyperplanes constructed by the machine. The optimal hyperplanes provide maximum separation (margin) between the data points (Boser et al. [1992], Cortes and Vapnik [1995]).

Core Algorithms: The error rate of an SVM machine on the test data (also called generalization) depends on the accuracy in learning a particular training set and its *capacity*, i.e., its ability to learn any training set without error. All SVM algorithms are designed such that there are zero errors in learning a particular training set, and the capacity is optimized, i.e., errors in learning any training set are minimized. It has been proven that optimal hyperplanes that provide maximum separation between data points in a high dimensional space lead to improved generalization (Vapnik [1998]).

Further, to construct such optimal hyperplanes, one needs to use only a subset of training data, called the *support vectors*.

Consider the linearly separable binary classification scenario. We have l training points, where each input x_i has d features and is in one of the two classes y_i = -1 or +1, i.e., the training set has the following form: $\{x_i, y_i\}, i = 1, \cdots, l, y_i \in \{1, -1\}, x \in R^d$. Since the data is linearly separable, one can partition the data into 2 classes using a hyperplane $x \cdot \mathbf{w} + b = 0$, where w is normal to the hyperplane and $\frac{b}{\|\mathbf{w}\|}$ is the perpendicular distance from the hyperplane to the origin. Let d_+ and d_- be the shortest distances from the separating hyperplanes to the closest training examples (i.e., support vectors). For the linearly separable case, the support vector algorithm simply looks for a separating hyperplane with the maximum margin, $d_+ + d_-$. This can be formulated as follows: suppose the training data set satisfy the following constraints:

$$x_i \cdot \mathbf{w} + b \geq +1, \text{ for } y_i = +1 \tag{2.4}$$

$$x_i \cdot \mathbf{w} + b \leq +1, \text{ for } y_i = -1. \tag{2.5}$$

These can be combined into one set of inequalities:

$$y_i(x_i \cdot \mathbf{w} + b) - 1 \geq 0 \ \forall i. \tag{2.6}$$

The points (support vectors) for which the equalities in Equations 2.4 and 2.5 hold, lie on two hyperplanes with $d_+ = d_- = \frac{1}{w}$ and the margin is $\frac{2}{\|\mathbf{w}\|}$. Maximizing the margin to the constraint in Equation 2.6 is equivalent to finding:

$$\min \| \mathbf{w} \| \quad s.t. \quad y_i(x_i \cdot \mathbf{w} + b) - 1 \geq 0 \ \forall i. \tag{2.7}$$

Minimizing $\| \mathbf{w} \|$ is equivalent to minimizing $\frac{1}{2}\| \mathbf{w} \|^2$. Thus, Equation 2.6 can be rewritten as

$$\min \frac{1}{2}\| \mathbf{w} \|^2 \quad s.t. \quad y_i(\mathbf{w}^T \cdot \phi(x_i) + b) - 1 \geq 0 \ \forall i, \tag{2.8}$$

where ϕ is a function that maps training data x_i into a higher dimensional space. Equation 2.8 can be formulated using Lagrange multipliers (Fletcher [2009]) and solved using the Quadratic or Linear Programming (QP/LP) approach to compute values of $\mathbf{w}$ and b. Existing general-purpose QP algorithms like the quasi-Newton or primal-dual interior point methods are usually used for small sized problems. For larger problems, LP solvers based on simplex or interior-point methods can be used (Section 2.13).

For data that is not linearly separable, a kernel function, $k(x_i, x_j)$, is used for mapping it to higher dimensional space. Most current SVM algorithms use one of the following basic kernel functions:

- Linear: $K(x_i, x_j) = x_i^T x_j$

- Polynomial: $K(x_i, x_j) = (\gamma + x_i \cdot x_j + r)^d, \gamma > 0$

- Radial Basis Function (RBF): $K(x_i, x_j) = \exp(-\gamma \parallel x_i - x_j \parallel^2), \gamma > 0$

- Sigmoid: $K(x_i, x_j) = \tanh(\gamma x_i^T x_j + r)$

Further Reading
Basic idea: Bennett and Campbell [2000], Boser et al. [1992], Burges [1998], Cortes and Vapnik [1995], Vapnik [1995, 1998]; **Algorithms:** Lodhi et al. [2002], Loosli and Canu [2007], Tsang et al. [2005]; **Packages:** R (The R Foundation), Weka (Hall et al. [2009]), IBM SPSS (IBM SPSS [2010a]), RapidMiner (Rapid-i), STATISTICA (The StatSoft Inc.), Oracle Data Miner (Oracle Corp. [b]), and SAS (SAS Institute Inc.); **Applications:** Guyon [2006], Hsu et al. [2010], Lodhi et al. [2002], Noble [2006], Rao et al. [2011], Sewell.

2.8   NEURAL NETWORKS

Artificial neural network is a system inspired by the biological network of neurons in the brain and uses a mathematical or computational model for information processing based on a connectionistic approach. A neural network can be viewed as a massively parallel distributed system that has a natural propensity for storing knowledge and mimics the brain in two respects: a neural network acquires knowledge through a learning process and stores it using inter-neuron connection strengths, represented using synaptic weights (Sarle [2002], Stergiou and Siganos).

Neural networks can be considered as nonlinear statistical data modeling or decision-making tools. In practice, they are used to model complex relationships between system input and output to infer results for novel inputs or finds patterns in data. Broadly, tasks to which neural networks are applied can be classified into: function approximation, classification, data mining, inferencing, and cognitive modeling. Neural networks have been applied to a diverse set of domains e.g., games (e.g., Go, Chess), music, material science, weather forecasting, medicine, chemistry, pattern recognitioni and classification (e.g., image, speech, or character), financial industry (e.g., analyzing stock trends), online fraud detection, and many more (Kriesel [2007], Sarle [2002], Widrow et al. [1994]).

Basic Idea: The most common neural network designs are based on the biological systems. In a biological network, neurons are linked to each other via weighted edges and when stimulated, they electrically transmit their signals via connecting axons. These signals get modified before reaching the destination neuron. A neuron gets multiple inputs that have been pre-processed and accumulated into a single pulse. A neuron, upon stimulation, may or may not emit a pulse. The output may be nonlinear and may not be proportional to the accumulated input (Kriesel [2007]).

Thus, a neural network implementation assumes that a neuron receives a *vector* input, $\vec{x}$ that is *weighted* and *accumulated* to a *scalar* value as a weighted sum before transmitting it to the receiver neuron ($\sum_i w_i x_i$). The weighted sum is an example of the *propagation* function. The set of such weights represent information storage of a neural network. The output of a neuron may not be proportional to the input (i.e., the response y is *nonlinear*, $y = f(\sum_i w_i x_i)$). The neural output is determined by its *activation* function $f()$. Multiple scalar output from different neurons in turn form the vector input of another neuron. Finally, the weights used in weighting the inputs are variables that capture the chemical processes in neurons.

The neural networks can be classified by: (1) underlying network topology, (2) type of learning algorithm used, and (3) type of input data. There are three major kinds of network topologies.

1. Feedforward networks: Feedforward networks consist of layers of neurons with connections to any one of the next layers. The neurons are grouped into the following layers: input layer, *n hidden* layers (invisible from outside), and output layer. These connections do not form any cycles.

2. Feedback networks: In feedback or recurrent networks, the state of a neuron at one time can influence its state at a future time. Some feedback networks allow direct cycles, in which a neuron is connected to itself; others only allow indirect cycles, where a neuron A acts as in input to neuron B and, in turn, neuron B is one of neuron A.

3. Completely linked networks: Completely linked networks permit connections between all neurons, except for direct recurrences. Furthermore, the connections need to be symmetric.

Neural networks are characterized by their capability to familiarize with problems by means of training and after sufficient training, to be able to solve unknown problems of the same class Kriesel [2007]. Neural networks learn by using a set of training patterns and modify their connecting weights as per certain rules (e.g., the Hebbian Rule; Hebb [1949]). There are three main types of learning schemes.

1. Unsupervised learning: In this approach, the training set consist of input patterns and the neural network tries by itself to detect similar patterns and classify them into pattern classes.

2. Reinforcement learning: In reinforcement learning, after completion of a training sequence, the network receives a response that specifies whether the result was right or wrong, if possible, *how* right or wrong it was.

3. Supervised learning: In supervised learning, the training set consists of input patterns and their correct results in form of activation from all output neurons. The objective is to change the weights so that not only the outputs match the values in the training set, but for unknown, similar patterns, the network produces plausible results.

Finally, neural networks are characterized by the kinds of input data. The most common kinds of data are categorical and quantitative. The categorical variables take only a finite number of possible values in different classes or categories, while the quantitative variables represent numerical measurements of some attributes. Learning with categorical values can be viewed as *classification*, while supervised learning with quantitative values is viewed as *regression*.

The most common form of neural network is called a perceptron, which is a feed-forward network with one input neuron layer connected to one or more trainable weight layers. A single-level perceptron (SLP) is a perceptron with an input layer and only one trainable weight layer. Neurons in the weight layer use a variety of activation functions (e.g., binary threshold, hyperbolic tangent, or weighted sum). A SLP with binary output is considered a linear classifier that maps its input real-valued vector x to an output binary value $f(x)$. A perceptron with two or more trainable weight layers is called a multi-level perceptron (MLP). An n-stage perceptron has n variable weight layers and $n + 1$ neuron layers, the first layer being the input layer, and $n - 1$ hidden layers. The hidden layers usually have nonlinear differentiable functions, e.g., logistic, softmax, and gaussian. Multi-layer perceptrons are usually trained using a supervised learning algorithm called *back-propagation*. The back-propagation algorithm involves two steps: (1) propagation that involves forward propagation of input through the neural network and backward propagation of the output activations to generate the delta errors; and (2) use the delta errors and input values to calculate the gradient of error of the network. The gradient is then used in a simple stochastic gradient descent algorithm to find weights that minimize the errors. This algorithm is basically a generation of the delta rule that is used for the single-layer perceptrons. First, the derivative of the error function with respect to the network weights is calculated and then the weights are modified such that the error decreases. For this reason, backpropagation can only be applied to nodes with differentiable activation functions.

Other important classes of neural networks include recurrent networks in which connections between neurons form a directed graph. Such networks are able to influence themselves by means of *recurrents*, using network outputs from the following computation steps (Hopfield [1982]). Another example is the convolution neural networks (CNNs, Fukushima [2013]) which are biological-inspired variants of MLPs. CNNs mimic operations of visual cortex by exploiting spatially-local correlations (LeCun et al. [1998]). CNNs are usually organized into in layers of two types: a convolution layer and a pooling (sub-sampling) layer, which computes the max or average value of a particular feature over a region of input data. Convolution and pooling enable the neural network to train in a translation-invariant manner (LeCun and Bengio [1995]). Recently, CNNs have gained a lot of attention as they are used extensively in deep learning systems: usually, a multi-stage neural networks which use CNNs as first layers that are connected to a follow-on layers of traditional MLPs (Bengio [2009], Bengio et al. [2014], Schmidhuber [2014]). Deep learning systems perform multiple nonlinear feature transformations over multiple stages.

Further Reading
Basic ideas: Hebb [1949], Kriesel [2007]; **Algorithms:** RBF Networks: (Kriesel [2007]), Hopfield Networks (Hopfield [1982]), Elman and Jordan networks: (Elman [1990], Jordan [1986]), Kohonen networks: (Kohonen [1997], Kohonen and Honkela [2007], Sarle [2002]); **Packages:** R (The R Foundation), Weka (Hall et al. [2009]), IBM SPSS, InfoSphere DataMining (IBM Corp. [c], IBM SPSS [2010a]), RapidMiner (Rapid-i), STATISTICA (The StatSoft Inc.), and SAS (SAS Institute Inc.); **Applications:** Kriesel [2007], Sarle [2002], Stergiou and Siganos, Widrow et al. [1994].

2.9   DECISION TREE LEARNING

Decision tree learning covers a class of algorithms that use a tree-based model (usually referred to as a decision tree) to represent decisions and their possible consequences (Shih [2004]). Intuitively, a decision tree is an encoding of all possible outcomes for a given problem scenario annotated with their conditional probabilities. Important applications of these algorithms include marketing, fraud detection, medical diagnostics, agriculture, and manufacturing/production (Murthy [1998]). For example, decision trees are used to determine if a potential customer should get a loan or to finalize treatment for a cancer patient. In these scenarios, the decision trees are used for classification in which the classifier model is used to predict values of categorical variables (e.g., *yes* or *no*) or continuous variables (e.g., amount of money a particular customer is willing to spend).

Basic Idea: Data classification is a two-stage process: in the first stage, a classifier model is built using a supervised learning process, and in the second stage the model is used to predict decisions for the unknown user inputs (i.e., for inputs that are not used in the training phase). The supervised learning phase uses a set of training *class-labeled* samples, where each sample is a n-dimensional *feature* vector and is associated with the corresponding *class-label attribute*. The class-label attribute can either be categorical or continuous. The distinct values of the class-label attribute define a distinct partition or *class* of the data set. The result of the learning phase is a decision tree whose internal nodes represent conjunction of feature predicates and leaves represent classifications. After the model has been trained, it is tested for accuracy using a new set of testing samples. Once the model's accuracy has been validated, it is ready for general data sets.

Most algorithms build the decision trees top-down by iteratively splitting the data set using a feature attribute at each step as the splitting parameter. Thus, as one traverses down the tree, the partitioning becomes more refined. One of the key factors affecting the performance of any decision tree algorithm is the selection of relevant feature attributes. Some features may be statistically correlated, thus *redundant*, and only one of these features is used for splitting. Further, some features may be *irrelevant* and can be completely eliminated during the splitting process. The reduced set of attributes can then generate a probability distribution of classes that is very similar to the original data set.

Thus, it is very important to select the feature predicates that can generate the *best* partitioning of the class-labeled data set into classes. The ideal partitioning would create distinct classes, each one would contain vectors that have the same values for the feature attributes used in the data set splitting. Conceptually, the *best* partitioning matches the ideal scenario as close as possible. Most decision tree algorithms use heuristics called *attribute selection measures* or *splitting rules* to select the feature predicates. The most popular attribute selection measures are: *Information Gain*, *Information Gain Ratio*, and *Gini Index* (Han and Kamber [2006], Lim et al. [2000], Loh and Shih [1997]). We now discuss key decision tree algorithms.

ID3/C4.5: ID3 (Iterative Dichotomiser) and C4.5 are two decision tree algorithms proposed by J. Ross Quinlan that use the entropy-based attribute selection measures (Kotsiantis [2007], Quinlan [1986, 1993]). Both algorithms use a greedy approach to build a decision tree in a top-down recursive divide-and-conquer manner using a class-labeled training set. Both ID3 and C4.5 algorithms require three parameters: the training set, D, set of attributes of the training vectors, and the attribute selection heuristic. The ID3 algorithm uses the information gain measure for attribute selection, while the C4.5 algorithm uses the information gain ratio.

Both algorithms build the tree starting with the root node N, that represents the entire training dataset D. If all vectors in D fall in the same class, the node N is considered a leaf node, and the process terminates. Otherwise, the algorithm uses the chosen attribute measure to determine the attributes will be used for partitioning the dataset. The node N is labeled with the splitting criterion that serves as the partitioning test for that node. For every outcome of the splitting criterion, a branch is grown from the node N. The dataset gets partitioned as per the distinct values of the attribute. The process terminates when all attributes have been used for partitioning or the vectors fall into the same class. Decision trees generated by ID3 or C4.5 suffer from the problem of *over-fitting* of noisy or outliner data. To address this problem, the tree is pruned to remove the least reliable branches after the fully grown tree has been built (called the *postpruning*). The C4.5 algorithm uses an approach called *pessimistic pruning* (Han and Kamber [2006]) that uses the training dataset to determine the pruning strategy.

CART: Classification and Regression Trees (CART or C&RT; Breiman et al. [1984]) is a family of nonparametric recursive tree-building algorithms for predicting continuous dependent variables (*regression*) and categorical predictor variables (*classification*). Like ID3/C4.5, CART builds the tree top-down by recursively partitioning the dataset. However, unlike the C4.5 algorithm, CART builds binary trees.

While building the tree, the CART uses two different measures for identifying the splitting attribute. For regression problems, a least-squares deviation criteria is used, while for categorical variables, impurity measures like the Gini index are employed. Once the splitting attribute is identified, the dataset is partitioned into two groups. The process continues until the stopping conditions are satisfied. The CART algorithms also use a postpruning approach to manage the resultant tree size. CART uses *cost complexity* to determine which part of the tree needs to be pruned (Han and Kamber [2006], Wu et al. [2008]). This postpruning approach generates a set of pruned trees; eventually, the tree that minimizes the cost complexity is selected.

CHAID: The CHAID (CHI-square Automatic Interaction Detector) algorithm also uses a recursive tree-building process that partitions the dataset as it build the tree (Kass [1980], Neville [1999]). Unlike binary tree created by the CART algorithm, CHAID creates a wide tree with multiple branches. As the CHAID algorithm can represent multiple categories effectively, it has been widely applied for market segmentation analysis. The CHAID algorithm uses the Pearson *CHI*-square test as the splitting criterion for ordinal and categorical variables and for continuous variables it uses F-tests.

The CHAID algorithm first prepares the predictor variables by dividing continuous distributions into a number of categories (*binning*). Internally, the CHAID algorithm only uses categorical variables. Once the initial categories are determined, the algorithm uses the *CHI*-squared or F-tests to determine statistical independence/significance of the data (Bonferroni-adjusted p-value). This information is used for determining the number of branches at an internal tree node. If the significance level is below a certain threshold, the branches are merged. Alternatively, a branch is split into two. The process terminates when there are no more significant splits or merges. In the CHAID algorithm, the last split determines the partitioning of the input dataset. A version of CHAID, called the *exhaustive CHAID*, chooses a partitioning that corresponds to the most significant split.

Further Reading
Basic idea: Han and Kamber [2006], Lim et al. [2000], Loh and Shih [1997], Quinlan [1986, 1993]; **Algorithms:** ID3/C4.5 (Kotsiantis [2007], Quinlan [1986, 1993]), C&RT (Breiman et al. [1984], Han and Kamber [2006]), CHAID (Kass [1980], Neville [1999]), and QUEST (Lim et al. [2000], Loh and Shih [1997], Shih [2004]); **Packages:** R (The R Foundation), Weka (Hall et al. [2009]), IBM SPSS (IBM SPSS [2010a]), RapidMiner (Rapid-i), STATISTICA (The StatSoft Inc.), Oracle Data Miner (Oracle Corp. [b]), and SAS (SAS Institute Inc.); **Applications:** Han and Kamber [2006], Murthy [1998], Nisbet et al. [2009], Shmueli et al. [2010].

2.10 TIME SERIES PROCESSING

A time series is a sequence of observations reported according to the time of their outcome (Falk et al. [2006]). Examples of time series data are prevalent in everyday life: prices of commodities during a trading day, daily opening and closing of stock market indexes, hourly weather reports, utility consumption charts, etc. Other important application domains that use time series data include geology, economics, control systems, medical informatics, process engineering, and social sciences (Croarkin and Tobias [2011], Shumway and Stoffer [2010]). Study of the time series data is targeted to achieve one of the two goals: (1) understand the basic characteristics of the observed data (**Analysis**); and (2) fit a model to the observed data set and apply it for forecasting based on known past values (**Forecasting**). Both the goals require the pattern for the time series to be identified and modeled.

Basic Idea: Analyzing a time series is different that the traditional data analysis as the data is not generated *independently*, dispersion of data items varies in time, it is often governed by a trend, and it can have cyclic components (Falk et al. [2006]). Thus, statistical approaches that assume *independent* and *identically distributed* data do not apply for the time series. The time series data inherently exhibits temporal ordering. In a time series, observations taken closer in time are more related than observations taken further apart. Further, an observation at a given time can be potentially derived from past observations, rather than from future observations.

In general, a time series $y_1, \cdots, y_n$ can be viewed as a sequence of random variables y_t that individually can be decomposed into four components:

$$y_t = T_t + Z_t + S_t + R_t \quad t = 1, \cdots, n, \tag{2.9}$$

where T_t is a monotone function of t, called *trend*, Z_t and S_t reflect long- and short-term non-random cyclic influences, called *seasonality*, and R_t represents a random variable capturing errors (*noise*) from the ideal non-stochastic model $y_t = T_t + Z_t + S_t$ (Falk et al. [2006], StatSoft Inc. [2010]). Analysis of any time series data involves identifying underlying trends and seasonalities. Time series analysis can be carried out either in time or frequency domain.

Trend Analysis: In the time-domain, trend analysis identifies the trend component of a time series. If the error component in the time series is significant, then the data needs to be pre-processed, or *smoothed*. The smoothing process involves some form of local averaging of data such that the irregular components of the individual observations cancel each other out. The most common technique is the *moving average* smoothing that replaces each element of the series by either simple or weighted average of n surrounding elements, where n is the width of the smoothing window (ARIMA; Box and Jenkins [1976]). The smoothing process can use medians instead of means: medians can reduce the effects of the outliners, but in absence of outliners, it can produce *jagged* curves. Medians also does not allow weighting during the smoothing process. In cases where the random errors are dominant, the smoothing process can use *distance weighted least squares smoothing* or *negative exponentially weighted smoothing* techniques (Falk et al. [2006], StatSoft Inc. [2010]). Once the errors are smoothed, the monotonous (increasing or decreasing) trend component of time series can be represented using linear or nonlinear functions, e.g., using the *logistic* function.

Seasonality Analysis: The seasonality component of the time series data captures the cyclic fluctuations in the data. In the time domain, the seasonality can be measured by evaluating dependences between elements of a time series separated with a distance or *lag k*. In the time-domain analysis, auto-correlation and auto-covariances are most commonly used as measures of dependence between time series elements. High values of auto-correlation at lag positions that are multiples of k exposes a pattern that repeats after every k elements. Auto-correlation values for consecutive lags are inter-dependent, i.e., they suffer from serial dependencies; if the first element is closely related to the second, and the second to the third, then the first element is related to the third element. One way to examine the serial dependencies is to use partial auto-correlation function that excludes all elements within the lag while calculating auto-correlation values (Box and Jenkins [1976]). The partial auto-correlation calculations for a lag of 1 are equivalent to computing auto-correlation. The serial dependencies within a time series for a lag of value k can be eliminated by differencing the series for the value k, i.e., replacing the ith element with its difference from the $(i - k)^{th}$ element. This transformation can reveal hidden seasonal characteristics by eliminating serial inter-dependencies. Secondly, the elimination of the serial dependencies can make the time series *stationary*, i.e., it has constant mean, variance, and auto-correlation over time (StatSoft Inc. [2010]).

Spectral Analysis: A time series can be viewed as a sum of a variety of cyclic components. These cyclic components are characterized using their wave-lengths as expressed via *periods* and *frequencies*. The frequency-domain (*spectral*) analysis of a time series aims to decompose the original time series into its cyclic components and to compute their frequencies to study their impact on the observed data. The spectral analysis uses two periodic sinusoidal functions, sine and cosine, to represent the original time series. This representation can be cast as a linear multiple regression problem, where the dependent variable is the observed time series, and the regression coefficients express the degree to which the respective sine and cosine functions are correlated with the data. A n-element time series will be represented by $\frac{n}{2} + 1$ cosine functions and $\frac{n}{2} - 1$ sine functions. Thus, an n-element time series will have n sinusoidal waves. The spectral analysis will identify correlations of sine and cosine functions of different frequencies with the observed data. If a large coefficient is found, there is strong correlation of the observed data with the corresponding frequency (i.e., an influential cycle with that frequency has been found).

Computationally, the spectral decomposition and identification of sine and cosine coefficients can be done using Fourier Transformations. For a n-element time series, the computations involve $O(n^2)$ complex operations. In practice, this process is implemented using the Fast Fourier Transform (FFT) algorithm which requires O(n lg(n)) operations.

Further Reading
Basic idea: Box and Jenkins [1976], Croarkin and Tobias [2011], Falk et al. [2006], StatSoft Inc. [2010]; **Algorithms:** ARIMA (Box and Jenkins [1976]), Exponential Smooth-

ing (Croarkin and Tobias [2011], StatSoft Inc. [2010]); **Packages:** R (The R Foundation), Weka (Hall et al. [2009]), IBM SPSS (IBM SPSS [2010a]), RapidMiner (Rapid-i), STATISTICA (The StatSoft Inc.), and SAS (SAS Institute Inc.); **Applications:** Nisbet et al. [2009], Shmueli et al. [2010], Shumway and Stoffer [2010].

2.11 TEXT ANALYTICS

Text analytics covers computational approaches that process structured and unstructured text data to extract and present innate information. The text analytics approaches usually operate on a *corpus* of text documents, potentially written in different languages, to transform the input data into a form that can be consumed by an user, usually a human subject. The goals of the text mining methods are to derive *new* information from data, find patterns across datasets, and separate relevant contextual information from noise. Text analytics is an inter-disciplinary field that uses techniques from statistics, natural language processing, linguistics, artificial intelligence, information retrieval, and data mining to pre-process, categorize, classify, and summarize the input text data.

One encounters text analytics extensively in daily life: from web searches and navigation, reading personalized online news articles, identification and filtering of e-mail spams, help desk communications, finding relevant references during research work, online advertisements, etc. Text analytics has been applied to a diverse class of domains which include intelligence gathering, bio-informatics, news gathering and classification, online advertising, medical informatics, social sciences, marketing, patent searching, and web searching/navigation. Common operations performed by text analytics tasks include pattern matching, lexical analysis, semantic analysis (e.g., synonym identification), entity recognition, co-reference, topic-bases classification, correlation, and document clustering, link analysis, and tagging/annotation. These capabilities are provided in most of the commercial text analytics packages (Davi et al. [2005], Feinerer et al. [2008], IBM SPSS [2010c]).

Basic Idea: The text analytics applications usually take text in its native raw format as input. The text can be organized as either a collection of documents or as individual text files. The input text can have imperfections like formatting, grammatical and typographical errors or contain unimportant stop-words like an or the. Thus, text analytics methods first pre-process the input data to clean it and prepare it for further analysis. Common steps in text pre-processing include (Feinerer et al. [2008]): import and parsing, stemming, whitespace elimination and case conversion, stopword removal, synonym identification, tagging, and annotations.

After the pre-processing stage, data from the input corpus is usually represented using specialized data structures. The most common text analytics data structure is the *term-document* matrix. This approach represents the processed text as a *bag of words* in which the order of tokens is irrelevant. The term-document matrix uses document IDs as rows and terms (tokens) as columns.

The matrix element (i, j) represent different weightings of a term j for the document i. Common weightings include term frequencies in a document, binary frequencies to represent inclusion or exclusion of a term, and inverse document frequency weighting that gives more weight to less frequent terms (more widely referred as TF/IDF). The term-document matrix is usually sparse and processed in compressed format. Alternatively, the pre-processed text is maintained in the native form and processed as strings (Lodhi et al. [2002]).

We now provide an overview typical applications of text mining that include natural language modeling, text categorization, text classification, semantic analysis, sentiment, and topic analysis.

- **Natural Language Modeling:** The first step in many text analytics workloads often involves understanding and processing input in specified in natural language such as English or Spanish. The core step in understanding natural language text is to build a language model (or an algorithm) that captures salient statistical characteristics of the distribution of sequence of words (Bengio et al. [2014]). The most common approach for building a language model uses a neural network to build a *distributed representation* of a word. The distributed representation represents a word using a vector of potentially non-mutual features that characterize the meaning of the word, where each vector entry captures the contribution of a feature on the word's meaning. The key approaches for learning word vectors include matrix-factorization based Latent Semantic Analysis (LSA) (Deerwester et al. [1990]), local context window approach such as the skip-gram model used in the word2vec system (Mikolov et al. [2013]), and exploiting word-word co-occurrences in the GloVe (Pennington et al. [2014]). The vector representations can then used for follow-on analysis such as similarity, clustering, etc.

- **Text Clustering:** Clustering allows (semi)-automatic categorization of text documents according to certain similarity measure (Conrad et al. [2005], Zhao and Karypis [2005b]). The sparse term-document representation of the text documents can be viewed as a representation of the data corpus in a high-dimensional space. The text data can then be clustered using traditional clustering algorithms like hierarchical clustering (Zhao and Karypis [2005a]) and k-means clustering (Section 2.3). Common similarity measures used for text clustering include metric distance, cosine distance, Pearson Correlation, and Extended Jaccard similarities (Strehl et al. [2000]).

- **Text Classification:** In contrast to clustering, text classification organizes text documents into pre-defined classes (Sebastiani [2002]). The class of a document is determined using document attributes, e.g., words. One of the most popular method of document classification is the Naive Bayes classifier. The Naive Bayes classifier assumes that all attributes of the training example are independent and generates training set to estimate parameters of the classification function. Alternative approaches include using support vector machines (SVMs) to classify documents (Joachims [1998], Lodhi et al. [2002]). SVMs are ideal for classifying data in a very high-dimensional feature space, e.g., when substrings are used as features, SVMs with string kernel functions can be used for classify documents such that documents with more common substrings in common are assigned to the same class.

- **Semantic Analysis:** Given a corpus of text documents, semantic analysis aims to extract and represent the innate semantic meaning, as approximated via contextual-usage of words, using either statistical or matrix computations. The most common technique used for semantic analysis, latent semantic analysis/indexing (LSA), uses matrix representation of the underlying text documents to analyze relationships between documents and words they contain, to infer deeper relationships between words, words and passages, etc. (Landauer and Dumais [2008]). The LSA creates a low-rank approximation of the term-document matrix using the Singular Value Decomposition (SVD) in which the k largest singular values are retained. The rank-reduction approximation eliminates any noise in the original data and merges dimensions associated with the words of similar meanings. The generated reduced-dimensional matrices can then be used for determining various similarities such as word-word or word-passage similarities.

- **Sentiment and Topic Analysis:** Sentiment analysis (opinion mining) aims to discover the *tone* of a document or sentence (e.g., positive, negative, or neutral) by applying natural language processing to the text data (Lakkaraju et al. [2011], Pang et al. [2002]). Sentiment analysis is related to topic analysis which aims to identify *hot* topics from the set of input text documents. Topic analysis uses various transformations of the term-document matrix, in particular, the non-negative matrix factorization (Lee and Seung [1999]), to deduce important topics. The non-negative matrix factorization is a family of unsupervised learning algorithms that view an object using parts-based additive representations. For topic analysis, NNMF can be used to explain a document as a linear combination of topics with non-negative weights; higher weights reflecting those topics that strongly influence the document.

Further Reading

Basic idea: Feinerer et al. [2008], Manning et al. [2009]; **Algorithms:** Naive Bayes (Manning et al. [2009], McCallum and Nigam [1998]), Latent Semantic Analysis (Deerwester et al. [1990], Landauer and Dumais [2008], Landauer et al. [1998]), String Kernel Functions (Lodhi et al. [2002]), and Non-negative Matrix Factorization (Ding et al. [2006], Ho [2008], Kanjani [2007], Lee and Seung [1999, 2001], Xu et al. [2003]); **Packages:** R (Feinerer et al. [2008]), SPSS Modeler (IBM SPSS [2010c]), STATISTICA (The StatSoft Inc.), Rapid-Miner (Rapid-i), Oracle Data Miner (Oracle Corp. [b]), and SAS (Davi et al. [2005]); **Applications:** Conrad et al. [2005], Gartner [2003], Joachims [1998,?], Lodhi et al. [2002], Pang et al. [2002], Sahami et al. [1998], Zhao and Karypis [2005b].

2.12   MONTE CARLO METHODS

The Monte Carlo method refers to a class of algorithms that employ repeated statistical sampling to compute *approximate* solutions to quantitative problems. These techniques are widely used for the applications with inherent uncertainty, such as pricing of various financial instruments, or for simulating systems with many coupled degrees of freedom, e.g., simulating behaviors of different materials. While the Monte Carlo methods had been originally designed for solving problems with probabilistic outcomes, they have also been applied to solve deterministic problems with infeasible computational requirements, e.g., solving multi-dimensional definite integrals with a large number of dimensions or with difficult boundary conditions.

The first application of the Monte Carlo methods was to understand behavior of nuclear reactions (e.g., neutrino travel patterns) (Metropolis and Ulam [1949]). Over the years, the Monte Carlo methodology has been applied to a wide array of applications domains including different physical sciences, engineering, finance, numerical analysis, and mathematical optimization (Fishman [1996]). Perhaps the most popular application of Monte Carlo methods is in financial engineering where they are extensively used for insurance risk modeling, pricing various types of options and derivatives, e.g., European and American style options, mortgage-backed securities, and portfolio analysis (e.g., calculating Value-At-Risk (VaR)) (Boyle [1977], Glasserman [2003]).

Basic Idea: The Monte Carlo method is defined as an approach that represents an approximate solution of a problem as a *parameter* of a hypothetical population. It uses a random sequence of values to construct a sample of the population, from which statistical estimates of the parameter can be obtained (Halton [1970]). For a problem, the Monte Carlo method repeatedly generates independent identically distributed random variables from the same distribution as the problem, and then uses them in either deterministic or stochastic model to compute the solution. In its simplest formulation, the Monte Carlo method can be viewed as an approach that uses statistical sampling to estimate a numerical integral. The standard error of a Monte Carlo estimation decreases with the square root of the sample size. Secondly, the standard error is independent of the dimensionality of the integral. Unlike the conventional numerical integration approaches which suffer from the curse of dimensionality, the amount of work required by the Monte Carlo approach does not increase exponentially in the number of dimensions. Unfortunately, to improve the estimation quality, the Monte Carlo approach usually requires a large number of samples. One way to improve the efficacy of the Monte Carlo methods is to use one of the variance reduction methods, e.g., importance or stratified sampling.

Sawilowsky [2003] classifies applications of Monte Carlo methods into three types: (1) using stochastic techniques for solving deterministic problems, e.g., Monte Carlo integration for solving multi-dimensional integral problems; (2) using stochastic techniques for solving problems with probabilistic outcomes (e.g., pricing of financial instruments), usually referred to as the *Monte Carlo Simulation*; and (3) using the Monte Carlo method as a tool for generating samples of a particular probabilistic distribution. However, in many practical scenarios, this distinction gets blurred. Algorithms that imply the Monte Carlo methods use the following methodology.

1. Identify a probability distribution function that mimics the problem under consideration (e.g., normal distribution for option pricing).

2. Generate samples from the probability distribution function using a pseudo-random number generator.

3. Pass the sample values through a deterministic or stochastic models (for simulation and sampling) to get the final result.

Irrespective of how the Monte Carlo methods are used in practice, all Monte Carlo implementations rely on techniques to generate *good* random number generators. In practice, it is not possible to generate *pure* random numbers, so Monte Carlo algorithms use either pseudo-random (PRNG) or quasi-random (QRNG) number generators. Most PRNG algorithms use bit manipulation and shuffling (e.g., the multiply-with-carry, xorshift, subtract-with-borrow, etc.) combined with recurrence strategies to incorporate pseudo-randomness into the generated sequences. Key PRNG algorithms include the Mersenne Twister (MT) generator Matsumoto and Nishimura [1998], and the Multiply-with-Carry algorithm (Marsaglia and Zaman [1991]). The most widely used quasi-random number generator is the Sobol sequence generator, which generates uniformly distributed numbers in the specified set of dimensions.

In practice, different variants of the original Monte Carlo algorithm are also used. One popular variant, the Markov Chain Monte Carlo (MCMC) (Metropolis et al. [1953]), is used for generating samples from a probability distribution by using a Markov Chain whose stationery distribution is the desired distribution. The MCMC algorithm is used for solving multi-dimensional integral problems and it has been also applied to scenarios that exhibit *random walk* behavior, e.g., in statistical physics or graphics applications. Another version, the Quasi-Monte Carlo method, uses low-discrepancy samples that are deterministically chosen based on equi-distributed sequences, i.e., appear to fill a region of n-dimensional space evenly. The low-discrepancy samples often lead to faster solution time and/or higher accuracy. The Quasi-Monte Carlo methods are used for solving multi-dimensional integration problems, e.g., pricing financial derivatives like Collateralized Mortgage Obligation (CDOs) (Peskov and Traub [1995]).

Further Reading
Basic idea: Halton [1970], Metropolis et al. [1953], Metropolis and Ulam [1949], Sawilowsky [2003]; **Algorithms:** Couture and L'Ecuyer [1994], Marsaglia and Zaman [1991], Matsumoto and Nishimura [1998, 2000], Saito and Matsumoto [2008]; **Applications:** Boyle [1977], Fishman [1996], Glasserman [2003], Halton [1970], Metropolis and Ulam [1949], Peskov and Traub [1995].

2.13   MATHEMATICAL PROGRAMMING

In a mathematical optimization or programming problem, one seeks to find an *optimal* solution for a problem scenario as defined by its constraints using a mathematical formulation. Specifically, solution of a mathematical programming problem aims to minimize or maximize a real *objective* function of real or integer variables, subject to constraints on the variables (Greenberg [2010], Holder [2006–2008]). The mathematical programming approach is usually applied to cases where a closed-form solution is not (easily) found and one has to settle for the *best available* solution. It forms the cornerstone of methods used in operations research and other related disciplines like industrial engineering, social sciences, economics, and management sciences. It has also been applied in a wide variety of domains including scheduling problems (e.g., transportation), manufacturing (e.g., steel production Dutta and Fourer [Fall 2001]), supply-chain management, product portfolio optimizations, workforce management, and product configuration selection.

Basic Idea: A *mathematical program* is an optimization problem of the form

$$Maximize \; f(x) : x \in X, \; g(x) \leq 0, \; h(x) = 0, \tag{2.10}$$

where X is a subset of R^n and is in the domain of the functions, f, g, and h, which map into real spaces. The relations, $x \in X$, $g(x) \leq 0$, and $h(x) = 0$ are called *constraints*, and the function f is called the *objective* or *cost* function Holder [2006–2008]. The domain X of the objective function is called the *search* space. A point x is *feasible* if $x \in X$ and it satisfies the constraints: $g(x) \leq 0$ and $h(x) = 0$. A point $x*$ is *optimal* if it is feasible and if the value of the objective function is not less than that of any other feasible solution: $f(x*) \geq f(x)$, for all feasible x (also called *candidate* or feasible solutions). This description uses *maximization* as the sense of optimization. The problem could be easily restated as a *minimization* problem by appropriately changing the meaning of the optimal solution: $f(x*) \leq f(x)$, for all feasible x.

The mathematical programming broadly covers approaches used for solving and using mathematical programs. It includes theorems to govern the form of the solutions, algorithms to seek a solution or ascertain none exists, formulations of problems as mathematical programs and theorems about quality of results, etc. In practice, mathematical programming approaches are classified according to the properties of the objective function, constraints, and candidate solutions. We now discuss important classes of mathematical programming:

Linear Programming: *Linear* programming approach is a special case of *convex* programming in which the the object function f is both linear and convex and the associated set of constraints are specified using only linear equalities or inequalities. Linear programming is used for modeling problems from operations research such as network flow, production planning, financial management, etc. Canonically, the linear programs can be expressed in matrix form as

$$Maximize \; c^T x \quad subject \; to \quad Ax \leq b \quad x \geq 0, \tag{2.11}$$

where x represents the vector of variables to be determined, c and b are vectors of known coefficients and A is a known matrix of coefficients. The set of constraints, $Ax \leq b$, form a convex polytope and any linear programming method would traverse over its vertices to find a point where the function, $c^T x$, has the maximum (or minimum) value, if such point exists. Most approaches for solving the linear programming problems explore the feasible region over a convex polytope defined by the linear constraints of the problem. The simplex method (Dantzig [1963]) is one of earliest, but still widely used, methods for solving LP problems. The simplex method constructs a feasible solution at a vertex of the polytope and then tests adjacent vertices by traversing a path on the edges of the polytope such that the objective function is improved or is unchanged. The simplex algorithm is very efficient in practice, requiring $2n$ to $3n$ iterations, where n is the number of equality constraints and is known to run in polynomial time in certain random inputs. The worst-case complexity of the simplex algorithm is exponential in the problem size. This problem was addressed by the Ellipsoid method (Schrijver [1998], Todd [2002]) that uses an iterative approach to generate a sequence of ellipsoids whose volumes uniformly reduce at every step, thus enclosing a minimizer of the convex objective function. The Ellipsoid method was the first solution to the linear programming problems that ran in worst-case linear time. The Karmarkar's algorithm (Karmarkar [1984], Todd [2002]) that uses an interior-point projection method improves on the Ellipsoid method's complexity bounds and runs in polynomial time for both average and worst cases.

Integer Programming: *Integer* programming (IP) family covers the set of linear programming applications that require values of the unknown variables to be integer. If only some of the variables are required to be integers, the problems are called *mixed-integer* programming (MIP) problems. Another version of the integer programming problem, 0-1 or *binary integer* programming, requires values of the unknown variables to be either 0 or 1. Integer programming problems often observed in scheduling scenarios, e.g., house building with workers with specific skills (IBM Corp. [b]) and in assignment-related problems, e.g., airline fleet assignment for optimal utilization or profit maximization (Abara [1989]). Integer programming is also used in a variety of distribution or network flow optimization problems.

Multiple variants of the IP problems are generally NP-Hard and usually solved using two classes of heuristics: cutting planes and branch-and-bound.

Combinatorial Programming: Combinatorial programming (or optimizations) cover methods that aim to optimize a cost function based on selection of objects from a set of objects (Papadimitriou and Steiglitz [1998]). Let $N = \{1, \cdots, n\}$ be a set of objects and let $\{S_1, S_2, \cdots, S_n\}$ be a finite collection of subsets. These subsets are characterized by inclusion and exclusion of objects based on certain conditions. Let each subset, S_k, be associated with a cost function, $f(S_k)$. The combinatorial optimization problem aims to select the subset of objects so as to maximize(minimize) the cost function. This formulation can be viewed as a special case of integer programming, whose decision variables are binary valued: $x(i, k) = 1$ if the ith element is in the

k^{th} set, S_k, otherwise, $x(i,k) = 0$ (Holder [2006–2008]). In practice, combinatorial optimization techniques are applied to a wide array of problems such as optimizing vehicle routing, VLSI circuit design, oil/gas pipeline design, steel/paper manufacturing, and matching factories with markets via intermediate warehouses (i.e., the transshipment problem).

Unlike linear programming, the feasibility space of the combinatorial algorithms is not convex; one needs to determine a global optimal point from several possible local optimal solutions. Although most combinatorial programming approaches are either NP-Hard or NP-Complete, in practice, many of these approaches can be solved in reasonable time by either choosing alternative formulations or exploiting specific features of a problem to compute its exact results (Hoffman [2000]). While special cases of some problems can be solved in polynomial time, most common approaches for solving combinatorial problems use either *approximation* algorithms that find a solution that is *provably* close to the optimal in *polynomial* time or use *heuristics* that search the feasibility space to compute sub-optimal solutions.

Constraint Optimizations: Constraint optimization (satisfaction) is a family of optimization problems that has a constant objective function with a set of constraints that impose conditions on the solution variables. The constraint satisfaction problems (CSPs) are characterized by constraints that are defined over a finite domain (Apt [2003], Dechter [2003]). In practice, CSPs has been used for scheduling problems, circuit layout in VLSI chips, DNA sequencing, production planning, computer vision, and computer games (e.g., sudoku) (Cork Constraint Computation Centre, University College Cork [2011], Moore [2011b]). A solution to a constraint satisfaction problem is a set of variables that satisfy all of the specified constraints.

The most common approach to solve the CSP problem is searching through different possible alternative solutions. Unlike the generalized search algorithms used in AI, the CSP search algorithms can use heuristics that exploit the problem structure and also perform search in any order. Most CSP algorithms use depth-first search to traverse the search tree. Often, the traversals reach a node where a node cannot be updated as the domain is empty. In such cases, the search algorithms *backtrack* to the previous assignment (i.e., a node at a higher level in the search) and restart the search with a new assignment. One of most widely used CSP search algorithms is the A* algorithm (Moore [2011a], Nilsson [1980]). The A* algorithm uses the best-first strategy to choose the next search node to traverse.

Nonlinear Programming: The nonlinear programming (NLP) can be viewed as a generalization of different mathematical programming formulations. In an NLP formulation, either or both the objective function or the constraints can be nonlinear. An NLP problem has the following form: minimize $f(x)$, subject to $g_i(x) = 0$, for $i = 1, \ldots, m_1$ and $h_j(x) \geq 0$, for $j = m_1 + 1, \ldots, m$, $m \geq m_1 \geq 0$. Depending on the nonlinearity, multiple special cases arise. Foe example, a scenario where the objective function is nonlinear, but the constraint functions g and h are linear, is called a *linearly constrained optimization*. If the objective function and the constraints are linear, an NLP problem reduces to a *linear programming* problem. When only the objective function is *quadratic*, the problem is termed *quadratic programming*. If the objective and constraint functions are defined over a convex set, the problem is called *convex optimizations* (Boyd and Vandenberghe [2009]). Finally, when both the objective function and constrains are nonlinear, the problem is called *unconstrained optimization* (Network-Enabled Optimization Systems Wiki).

The difficulty in solving an NLP problem arises from the fact that NLP problems have non-convex object function or constraints. Such problems can have multiple feasible regions and multiple locally optimal points within each regions. Consequently, the non-convex NLP problems can exhibit solutions with *local optima*: they are spurious solutions that satisfy the requirements on the derivatives of the constraint functions. To determine if an NLP problem is infeasible (e.g., the objective function is unbounded) or a solution is the *global optimum* can require time *exponential* in the number of variables and constraints. The complexity of solving an NLP problem varies according to its type: in general, problems with convex objective function or constraints (e.g., quadratic programming) are easiest to solve, while problems that aim to find the *global* optimal are much harder to solve.

Further Reading
Basic idea: Greenberg [2010], Holder [2006–2008], Network-Enabled Optimization Systems Wiki; **Algorithms:** Linear Programming (Dantzig [1963], Gill et al. [1986], Marsten et al. [1990], Mehrotra [1992,?], Schrijver [1998], Todd [2002,?], Weisstein [2011]), Integer Programming (Barnhart et al. [1998]), **Combinatorial Programming** (Hoffman [2000], Kirkpatrick et al. [1983], Papadimitriou and Steiglitz [1998], Schrijver [2002], Spall [2003], Vazirani [2003]), **Constraint Optimizations** (Apt [2003], Dechter [2003], Moore [2011a], Nilsson [1980]), **Nonlinear Programming** (Boyd and Vandenberghe [2009], Gould and Toint, Network-Enabled Optimization Systems Wiki, Neumaier [2004]); **Packages:** IBM ILOG CPEX IBM Corp. [b], COIN-OR COIN-OR Foundation [2011], Gurobi Gurobi Optimization Inc.; **Applications:** Abara [1989], Anderson et al. [2009], Armacost et al. [2004], Cork Constraint Computation Centre, University College Cork [2011], Dutta and Fourer [Fall 2001], IBM Corp. [b], Moore [2011b], Papadimitriou and Steiglitz [1998], Schrijver [2002].

2.14   ON-LINE ANALYTICAL PROCESSING

On-line analytical processing or OLAP refers to a broad class of analytics techniques that process historical data using a logical multi-dimensional data model (Chaudhuri and Dayal [1997], Codd et al. [1993a,b]). Over the years, OLAP has emerged to become the key *business intelligence* (BI) technology for solving *decision support* problems like business reporting, financial planning and budgeting/forecasting, trend analysis and resource management. OLAP technologies usually operate on *data warehouses* which are collections of *subject-oriented, integrated, time-varying, non-*

volatile, and historical collection of data (Chaudhuri and Dayal [1997]). Unlike on-line transaction processing (OLTP) applications that support repetitive, short, atomic transactions, OLAP applications are targeted for processing complex and ad-hoc queries over very large (multi-Terabyte and more) historical data stored in the data warehouses.

Basic Idea: OLAP applications are targeted for knowledge workers (e.g., analysts, managers) who want to extract useful information from a set of large disparate data sources stored in the data warehouses. These sources can be semantically or structurally different and can contain historical data consolidated over long time periods. OLAP workloads involve queries that explore relationships within the underlying data and then exploit the acquired knowledge for different decision support activities such as post-mortem analysis/reporting, prediction, and forecasting. The OLAP queries tend to invoke complex operations (e.g., aggregations, grouping) over a large number of data items or records. Thus, unlike the OLTP workloads, where transaction throughput is important, query throughput and response times are more relevant for OLAP workloads. Thus, an OLAP system needs to support a logical model that can represent relationships between between records succinctly, a query system that can explore and exploit these relationships, and an implementation that can provide scalable performance.

Logical Data Model: Most OLAP systems are based on a logical data model that views data in the warehouse as multi-dimensional data *cubes*. The multi-dimensional data model grew out of the two-dimensional array-based data representation popularized by the spreadsheet applications used by business analysts (Chaudhuri and Dayal [1997], Gray et al. [1997], Harinarayan et al. [1996]). The data cube is typically organized around a central theme, e.g., car sales. This theme is usually captured using one or more *numeric measures* or *facts* that are the objects of analysis (e.g., number of cars sold and the sales amount in dollars). Other examples of numerical measures include budget, revenue, retail inventory. The measures are associated with a set of independent *dimensions* that provides the context. For example, the dimensions associated with the car sales measure can include the car brand, model and type, various car attributes (e.g., color), geography, and time. Each measure value is associated with an unique combination of the dimension values. Thus, a measure value can be viewed as an entry in a *cell* of a multi-dimensional *cube* with a specified number of dimensions.

In the multi-dimensional OLAP model, each dimension can be further characterized using a set of attributes, e.g., the geography dimension can consist of country, region, state, and city. The attributes can be viewed as sub-dimensions and can themselves be related in a hierarchical manner. The attribute hierarchy is a series of parent-child relationships that is specified by the order of attributes, e.g., year, month, week, and date. A dimension can be associated with more than one hierarchy. The parent-child relationship represents the order of summarization via aggregation: the measure values associated of a parent are computed via aggregation of measures of its children. Thus, the dimensions, along with their hierarchical attributes, and the corresponding measures, can be used to capture the relationships in the data.

OLAP Queries: Typical OLAP analytics queries perform two main functions: reporting and presentation. Reporting involves organizing dimensions and performing computations on the corresponding measures. Presentation involves selecting dimensions and measures from the original or computed versions of the data, and preparing them for display. Functionally, OLAP queries can be classified into *what-now* (post-mortem analysis), *what-if* (prediction), and *what-next* (forecasting). To support these analyses, an OLAP engine supports a number of operators. Some of the key operators include: *Group-by* which collates the measures as per the unique values of the specified dimensions, *slice_and_dice* which involves reducing the dimensionality of the dataset by taking a projection of the data on a subset of dimensions for selected values of the other dimensions, *pivoting* or rotating operation re-orients the original cube to visualize the data using new relationships, and *rollup* and *drill-down* operators support aggregation across hierarchies within one or more dimensions. OLAP analysis also involves invoking a variety of analytical functions on measure values. The OLAP analytical functions can be broadly classified into aggregation (e.g., sum), scalar, and set functions (e.g., sort). These functions can be either user-defined or pre-defined by the underlying system.

OLAP Servers Implementations In practice, the multi-dimensional OLAP model is usually implemented using one of the three approaches: Relational OLAP (ROLAP), Multi-dimensional OLAP (MOLAP), or Hybrid OLAP (HOLAP) (Chaudhuri and Dayal [1997]). In the ROLAP approach, a relational database system is used for storing and processing data in the multi-dimensional OLAP model. The data warehouse is implemented as relations stored in tables and queried using SQL-based OLAP queries.

The MOLAP approach stores and processes the multi-dimensional OLAP cubes as multi-dimensional arrays. In most cases, the MOLAP cubes are sparse multi-dimensional arrays that are stored using specialized data structures to optimize data access costs. Data stored in the MOLAP fashion is queried using languages that can express data access using the multi-dimensional array model, e.g., Microsoft's Multidimensional Expressions (MDX) language (Microsoft Developer Network). The hybrid OLAP strategy uses a combination of relational amd multi-dimensional OLAP implementations to store and process OLAP data. There are two ways for partitioning data between ROLAP and MOLAP stores: the first strategy stores the materialized view for a query workload in the MOLAP format and maintains the raw, detailed data in the ROLAP format, while the second stores some section of the data (e.g., most recent or most commonly used) in the MOLAP format, while maintaining the remaining data in the ROLAP format. Examples of systems that use the HOLAP approach include OLAP servers from Microsoft, Oracle, and SAP (Business Application Research Center).

Further Reading

Basic Idea: Chaudhuri and Dayal [1997], Codd et al. [1993a,b], Gray et al. [1997], **Algo-**

rithms: Chaudhuri and Dayal [1997], Gray et al. [1997], Harinarayan et al. [1996], Willhalm et al. [2009a]; **Packages:** Microsoft SQL (Microsoft Corp., Microsoft Developer Network), IBM DB2 and Cognos TM1 (Chamberlin [1998], IBM Corp. [a]), Oracle (Oracle Corp. [a]), Teradata (Teradata Inc.), HP Vertica (Vertica Systems Inc. [2010]), IBM Netezza (IBM Netezza), SAP (Intel Corp. [2011], SAP Inc. [2010]) and Palo (Jedox AG.); **OLAP Applications:** (Business Application Research Center, Chaudhuri and Dayal [1997]).

2.15 GRAPH ANALYTICS

Graphs and related data structures (e.g., trees and directed-acyclic graphs (DAGs)) form the fundamental tools used for expressing and analyzing relationships between entities. Relationships modeled by graphs include associations, hierarchies, sequences, positions, and paths (Barabasi [2003], Chakrabarti and Faloutsos [2006], Newman [2010]). Graphs have been used in a wide array of diverse application domains: from biology, chemistry, pharmacology, linguistics, economics, and operations research to different problems in computer science.

Basic Idea: Formally, a graph $G = (V, E)$ is described using a set of vertices (or nodes) V and the edges E that connect them. The graph vertices or edges can be weighted, and the edges can be directed or undirected. Graph vertices can also support additional attributes, e.g., a *color*. Usually, the graph is traversed in a pattern which is determined by either the graph characteristics or external constraints. This basic formulation of the graphs can be used for building complex data models. For example, molecular structure of chemical compounds is usually represented using graphs whose nodes represent atoms and edges represent bonds; a node- and edge-weighted graph can represent a transportation system between cities of a region, where the node weight represents city population and edge weight represents transportation density. Other applications of graphs include biological modeling, sociology (e.g., social network analysis), linguistics (e.g., expressing language syntax and semantics), electrical circuit design, combinatorial optimizations (e.g., flow problems), and neuroscience (e.g., modeling brain's cognitive connections) (van den Heuvel et al. [2008]).

Graph analytics refers to a class of techniques that either use graph models to solve a problem (e.g., the traveling salesman and other optimization problems), or to analyze and exploit inherent graph structures of a problem (e.g., identifying sub-graphs with a well-defined structure, *motifs* (Milo et al. [2002]), from graphs representing chemical compounds). Broadly, the graph algorithms can be classified into three overlapping categories: structural algorithms that analyze and exploit different topological properties of a graph, traversal algorithms that navigate different paths in a graph, and pattern-matching algorithms that find instances of different graph patterns (e.g., cycles) in a graph. These algorithms are characterized by how the graph data is interpreted and analyzed, and how the graphs are represented. Common graph representations include directed and undirected graphs, graphs with weights on the edges and vertices, rooted graphs, commonly called trees, and their variants. In practice, the graphs are represented using two different formulations: the first one enumerates the graph edges using lists, while the second uses matrices to capture the graph structure. The list-based data structures include adjacency list that colocates a vertex with its neighbors (i.e., vertices with direct connections), and incident list that lists all edges as pairs or tuples (for directed graphs). The list-based representation is popular in the algorithms that navigate the graph structure. The matrix formulation is used to provide a concise representing of different structural attributes of a graph, e.g., connectivity, weights, direction, etc. In most situations, the graph matrices are sparse and are implemented using compact array-based data structures.

Structural Algorithms: Graph structural algorithms, commonly known as network analysis algorithms, analyze symmetric and asymmetric relationships between networked entities by exploring structure of the underlying graph. Usually, the networked data is represented via digraphs or *networks* (Barabasi [2003], Chakrabarti and Faloutsos [2006], Newman [2010]). These networks can be structurally classified into two categories: small-world networks in which distance between any two randomly chosen vertices grows proportional to the logarithm of the total number of vertices in the network (Watts and Strogatz [1998]), and scale-free networks, where the degree distribution follows the power law (Barabasi and Bonabeau [2003]). Graph structural algorithms are designed to understand and exploit inherent abstract structural properties of a network. Such structural information can be used for different purposes, for example, telecom companies can use the structural information of the call graphs to identify customers most likely to switch carriers (also called churn analysis; Dasgupta et al. [2008], Nanavati et al. [2006], Richter et al. [2010]).

Traversal Algorithms: The second class of graph algorithms involves traversing edges of a graph to find solution of the associated problem. Graph traversal algorithms operate on graphs that either capture the structure of some underlying physical network (e.g., roads, pipes, etc.) or capture the abstract model of a problem (e.g., a tree representing an XML document, or a graph representing cities and distances in a traveling salesman problem). Unlike the structural algorithms where in many cases analytical solutions are computable via matrix formulation, problems addressed using traversal algorithms are notoriously complicated to solve—many of them are NP-Complete, and thus the algorithms must make extensive use of heuristics.

   The traversal algorithms are used to solve: (1) *route* problems which aim to optimize path lengths under different traversal constraints, (2) *flow* problems that investigate flow of material (e.g., oil, gas, cars, etc.) over a network that is represented by the underlying directed graph, (3) *coloring* problems that label graph elements (e.g., vertices) to satisfy certain constraints, and (4) *searching* problems that find a problem solution by traversing vertices which encode the problem states.

Pattern-matching Algorithms: The final class of graph algorithms focuses on finding different patterns in an input graph. Most common graph patterns include cycles, various types of cliques (an undirected graph formed using a subset of vertices such that every two vertices are connected), sub-graphs with certain properties (e.g., isomorphic with some other sub-graph), and network motifs (Chakrabarti and Faloutsos [2006]). Important practical applications of pattern matching include: social analytics, workforce analytics, epidemiology, financial network modeling, and neuroscience. Graph pattern matching also forms the basic tool for clustering nodes from a graph based on a similarity metric (Schaeffer [2007]).

The generalized combinatorial problems of enumerating or identifying structural patterns (e.g., finding maximal sub-graphs) in a graph are NP-Complete. Hence, these problems are solved approximately by using heuristics or their constrained versions are solved in polynomial time. While a majority of graph pattern-matching algorithms use traversal-based approaches to reach at the solution, some pattern-matching problems can be solved using the matrix representation (e.g., using adjacency matrix) of a graph. In particular, in certain scenarios, the clique discovery problem can be viewed as a graph partitioning or clustering problem and solved using spectral clustering techniques (Schaeffer [2007]), including the non-negative matrix factorization approach (Ding et al. [2008]).

Further Reading
Basic Idea: Chakrabarti and Faloutsos [2006], Harary [1969], Newman [2010]; Algorithms: Structural (Alon [1998], Barabasi and Bonabeau [2003], Dhillon et al. [2005], Page et al. [1999], Watts and Strogatz [1998]), Traversal (Consortium [a,b], Harary [1969]), and Pattern-matching (Baskerville and Paczuski [2006], Bron and Kerbosch [1973], Chakrabarti and Faloutsos [2006], Ding et al. [2008], Grochow and Kellis [2007], Schaeffer [2007]); **Packages:** SPSS Modeler (IBM SPSS [2010a]) and R (The R Foundation); **Applications:** Barabasi [2003], Cecchi et al. [2008], Chakrabarti and Faloutsos [2006], Dasgupta et al. [2008], Krebs, Leskovec et al. [2014], Lohmann et al. [2010], Micheloyannis et al. [2006], Milo et al. [2002], Nanavati et al. [2006], Podolyan and Karypis [2008], Porter et al. [2009], Richter et al. [2010], Stam et al. [2006], van den Heuvel et al. [2008].

CHAPTER 3

Accelerating Analytics

3.1 CHARACTERIZING ANALYTICS EXEMPLARS

In the previous chapters, we discussed various stages in a typical analytics execution workflow and presented a set of analytics model exemplars. In this chapter, we focus on computational and runtime characteristics of these exemplars and discuss its impact on accelerating various analytics algorithms.

Table 3.1 presents a summary of the analytics exemplars with their associated problem types, functional goals (Table 1.3), and key algorithms. As Table 3.1 illustrates, an analytics exemplar can have multiple functional goals. For example, regression algorithms can be used for either for predicting target classes for input data (classification) or as a statistical tool for quantitative analysis. The Naive Bayes algorithm can be used in text analytics or for general clustering purposes. Each exemplar can be implemented by multiple algorithms, each designed for specific runtime and user constraints (e.g., linear, logistic, and probit regression). Further, an analytics algorithm can be used by different exemplars for achieving different functional goals. For example, logistic regression can be used as a statistical tool in a data analysis workload or as a classifier in a neural network workload. Each algorithm, depending on the runtime constraints, i.e., whether the application data can fit into main memory or not, can use a variety of algorithmic kernels (Figure 1.2). Finally, a real analytics workload consists of one or more of the analytics components, each with potentially different functional goals, runtime requirements and data requirements. For example, key components of the IBM Watson DeepQA system used in the Jeopardy! Challenge include natural language processing of input queries, regression for ranking candidate answers, and simulation for modeling different waging scenarios (Ferrucci et al. [2010]).

Table 3.2 presents a summary of computational patterns, key data types, data structures, and functions used by algorithms for each exemplar, while Table 3.3 summarizes the runtime characteristics of these exemplars.

3.1.1 COMPUTATIONAL PATTERNS

As Table 3.2 illustrates, while different exemplars demonstrate distinct computational characteristics, they also exhibit key similarities. Most analytics exemplars operate on data that can be inherently structured or unstructured (notable exception being Monte Carlo methods that generate new data based on few input parameters). Algorithmically, many exemplars operate on sparse and high-dimensional data. Such data needs to be transformed so as to make the task computationally feasible, e.g., dimensionality reduction using principal component analysis or singular

Table 3.1: Analytics exemplar models, along with problem types and key application domains

Analytics Exemplar (Problem type)	Functional Goals	Key Algorithms
Regression analysis (Inferential statistics)	Prediction Quantitative Analysis	Linear, Nonlinear, Logistic Regression Probit Regression
Clustering (Supervised learning)	Recommendation, Prediction, Reporting	K-Means and Hierarchical clustering EM Clustering, Naive Bayes
Nearest-neighbor search (Unsupervised learning)	Prediction, Recommendation	K-d, Ball, and Metric trees Locality-sensitive Hashing Approx. Nearest-neighbor
Association rule mining (Unsupervised learning)	Recommendation	Apriori, Partition, FP-Growth, Eclat and MaxClique, Decision trees
Recommender Systems (Unsupervised learning)	Recommendation	Pearson Colleration, Latent factor/Non-negative fatorization Nearest-neighbor, Naive-Bayes Classifier
Neural networks (Supervised learning)	Prediction Pattern matching	Single- and Multi-level perceptrons, RBF, Recurrent, and Kohonen networks
Support Vector Machines (Supervised learning)	Prediction Pattern matching	SVMs with Linear, Polynomial, RBF, Sigmoid, and String kernels
Decision tree learning (Supervised learning)	Prediction Recommendation	ID3/C4.5, CART, CHAID, QUEST
Time series processing (Data analysis)	Pattern matching, Alerting	Trend, Seasonality, Spectral analysis, ARIMA, Exponential smoothing
Text analytics (Data analysis)	Pattern matching Reporting Prediction	Naive Bayes classifier, String-kernel SVMs, Latent semantic analysis Non-negative matrix factorization
Monte Carlo methods (Modeling and simulation)	Simulation Quantitative analysis	Markov-chain, Quasi-Monte Carlo methods
Mathematical programming (Optimization)	Prescription Quantitative analysis	Primal-dual interior point, Branch and Bound Methods A* algorithm, Quadratic Programming
On-line analytical processing (Structured data analysis)	Reporting Prediction	Group-By, Slice_and_Dice, Pivoting, Rollup and Drill-down, Cube
Graph analytics (Unstructured data analysis)	Pattern matching Recommendation	Eigenvector Centrality, Routing, Searching and flow algorithms Clique and motif finding

value decompostion (Berry et al. [1994]). Further, many exemplars are formulated as optimization problems which terminate when the target cost function is optimized.

These factors impact the data structure design, computational patterns, and functions of the exemplars. Broadly, the analytic exemplars can be classified into two classes: the first class exploits mathematical (e.g., linear algebraic) formulations and the second operates on non-numeric data structures. Exemplars belonging to the first class, e.g., mathematical programming, Monte

Table 3.2: Computational characteristics of the analytics exemplars

Analytics Exemplar	Computational Patterns	Data types, Data structures, and Functions
Regression Analysis	Matrix inversion, LU decomposition Transpose, Factorization	Double-precision and Complex data Sparse/Dense matrices, Vectors Dot Product Calculations
Clustering	Cost-based iterative convergence	Height-balanced tree, Graph, Distance functions, log function
Nearest-Neighbor Search	Distance calculations, Hashing Singular value decomposition,	Higher-dimensional data structures, Hash tables, Distance functions Dot Product Calculations
Association Rule Mining	Set intersections, Unions, and Counting	Hash-tree, Prefix trees, Bit vectors
Recommender Systems	Vector-space Similarity, Latent and non-negative factorization Naive-Bayes classifer, Nearest neighbor	Vectors, Sparse/Dense matrices Single-/Double-precision data Dot Product Calculations
Neural Networks	Matrix-Matrix/-Vector Multiplications Gradient Descent Algorithms FFTs, Convolution, Cross-correlation	Sparse/dense matrices, Vectors, Single-/Double-precision, Complex data, Dot Product
Support Vector Machines	Linear Solvers	Double-precision Sparse matrices Vectors, Kernel functions Dot Product Calculations
Decision Trees	Dynamic programming Recursive Tree Operations	Integers, Double-precision, Trees, Vectors, log function
Time Series Processing	Smoothing via averaging, Correlation Fourier and Wavelet transforms	Integers, Single-/Double-precision Dense matrices and Vectors Distance and Smoothing functions
Text Analytics	Parsing, Bayesian modeling Matrix Factorization, Multiplication Hashing, String Matching Set Operations (Union and Intersection)	Integers, Single-/Double-precision, Sparse matrices, Vectors, Strings, Distance functions String functions, Inverse Indexes Dot Product Calculations
Monte Carlo Methods	Random number generators Polynomial evaluation, Interpolation	Double-precision, Bit vectors Bit-level operationss
Mathematical Programming	Linear Solvers, Factorization Dynamic programming, Greedy Algorithms	Integers, Double-precision, Vectors, Trees, Sparse matrices, Adjacency list
On-line Analytical Processing	Grouping and ordering Aggregation over hierarchies	Prefix trees, Relational tables, Sorting, Ordering, OLAP Operators
Graph Analytics	Graph traversal, Eigensolvers, Matrix-Matrix/-Vector multiplication Non-negative matrix factorization	Integer, Single-/Double-precision Sparse matrices, Trees Adjacency Lists

Carlo methods, regression analysis, recommender Systems, support vector machines, and neural networks use linear-algebraic formulations to capture relationships in the underlying data. The second class, which includes clustering, nearest-neighbor search, associative rule mining, decision tree learning, and OLAP use non-matrix data structures. Exemplars like text analytics, and graph analytics can use either approaches. Linear algrebraic frameworks usually use two- or three-dimensional matrices to encode relationships in low-dimensional data and vectors for representing locations in high-dimensional data space. Based on the type of data relationship, matrices can be either sparse or dense and are used in various linear algebraic kernels like matrix-matrix, matrix-vector multiplications, inversion, transpose, linear solvers, and various factorizations such as Cholesky factorization, Singular Value Decomposition, and Non-negative matrix factorization (Berry et al. [1994], Kleinberg and Tomkins [1999]). In case the data has complex relationships (e.g., OLAP data organized in multiple shared hierarchies) or employs operations that cannot be encoded as matrix or vector operations, analytics exemplars use data structures like hash tables, queues, sets, graphs, inverse indexes, adjacency lists, and trees. Common operations on these data structures include traversals, hash table queries, set union and intersections, sorting, and grouping.

The analytic exemplars use a variety of types, such as integers, strings, bit vectors, and floating point variables (e.g., single, double precision, and complex) to represent input data and output results. Most exemplars require high precision calculations (certain mathematical programming algorithms require both high precision and result repeatability). One notable exception is neural networks, where certain algorithms can be implemented using low precision types (Gupta et al. [2015]).

Finally, these exemplars use a wide array of functions to compare, transform, and modify input data. Examples of common analytic functions include various distance functions (e.g., Euclidian), kernel functions (e.g., Linear, Sigmoid), statistical and aggregation functions (e.g., Sum, Min, or Average), data organization functions (e.g., sorting, hashing, and grouping), smoothing functions (e.g., correlation). These functions, in turn, make use of intrinsic library functions such as `log`, `sine`, or `sqrt` or various bit manipulation routines.

3.1.2 RUNTIME CHARACTERISTICS

Table 3.3 summarizes the runtime characteristics of the analytics exemplars. The key distinguishing feature of analytics applications is that they usually process input data in the read-mostly format. The input data can be scalar, structured (e.g., images, or unstructured (e.g., raw text), and is usually read from files, streams or relational tables in the binary or text format. In most cases, the input data is large, which requires analytics applications to store and process data from disks. In case of time-series processing, the large volume data is usually streamed, and can be both structured (e.g., web-service messages) or unstructured (e.g., text messages). Notable exceptions to this pattern are Monte Carlo Methods and Mathematical Programming, which are inherently in-memory as they operate on small input data. In most cases, the results of analysis are usually smaller than the input data. Only three exemplars—association rule mining, Monte Carlo methods, and on-line analytical processing—generate larger output. Most analytics exemplars, with the exception of time-series processing, operate in the batch mode; time-series processing has real-time constraints. Finally, analytics algorithms can involve one or more stages (e.g., in supervised learning), where each stage can invoke the underlying algorithm in an iterative or non-iterative manner. For the iterative workloads, for the same input data size, the running time can vary depending on the precision required in the results.

Table 3.3: Runtime characteristics of the analytics exemplars

Analytics Exemplar	Execution characteristics		Input-Output characteristics	
	Methodology	Memory Issues	Input Data	Output Data
Regression Analysis	Iterative	In-memory Disk-based	Large historical Structured	Small Scalar
Clustering	Iterative	In-memory Disk-based	Large historical Unstructured Structured	Small scalar Unstructured Structured
Nearest-Neighbor Search	Non-iterative	In-memory	Large historical Structured	Small Scalar Structured
Association Rule Mining	Iterative Non-iterative	In-memory Disk-based	Large historical Structured	Larger Structured
Recommendation Systems	Non-iterative	Disk-based	Large historical Structured	Small Structured
Neural Networks	Iterative Two Stages	In-memory Disk-based	Large Structured	Small Scalar
Support Vector Machines	Iterative Two Stages	In-memory Disk-based	Large Structured	Small Scalar
Decision Tree Learning	Iterative Two Stages	In-memory Disk-based	Large Unstructured	Small Scalar
Time Series Processing	Non-iterative Real-time	In-memory	High volume streaming Unstructured Structured	Smaller Scalar Streaming
Text Analytics	Iterative Non-iterative	In-memory Disk-based	Large historical Unstructured Structured	Large/small Unstructured Structured
Monte Carlo Methods	Iterative	In-memory	Small Scalar	Large Scalar
Mathematical Programming	Iterative	In-memory	Small Scalar	Small Scalar
On-line Analytical Processing (OLAP)	Non-iterative	In-memory Disk-based	Large historical Structured	Larger Structured
Graph Analytics	Iterative	In-memory Disk-based	Large historical Unstructured	Small Unstructured

3.2   IMPLICATIONS ON ACCELERATION

Given the varied computational and runtime characteristics of the analytics exemplars, it is clear that a single solution for accelerating different analytics applications would be sub-optimal. As Tables 3.2 and 3.3 demonstrate, each exemplar has a unique set of computational and runtime features, and ideally, every exemplar would get a system tailor-made to match its requirements. However, we have also observed that different analytic exemplars share many computational and runtime features. Therefore, for a systems designer, the challenge is to customize analytics systems using as many re-usable software and hardware components as possible. In this section, we describe various opportunities for accelerating analytics workloads on *existing* software and hardware systems, and then discuss how to build re-usable accelerated components.

3.2.1   SYSTEM ACCELERATION OPPORTUNITIES

Based on the computational and runtime characteristics described in Tables 3.2 and 3.3, we first classify the analytics exemplars based on their performance bottlenecks:

- Compute-bound Exemplars: Mathematical Programming and Monte Carlo Methods

- Compute-bound or Network-/Memory-bound Exemplar: Time-series Processing

- Compute-bound (when in-memory) and I/O-bound (when disk-based): Text Analytics, Regression Analysis, Clustering, Nearest-neighbor Search, Neural Networks, Support Vector Machines, Recommender Systems

- Memory-bound (when in-memory) and I/O-bound (when disk-based): OLAP, Graph Analytics, Text Analytics, Decision Tree Learning, Associated Rule Mining

Among all the exemplars, only two—mathematical programming and Monte Carlo methods—are purely compute-bound. A majority of the remaining exemplars are compute-bound when the data is entirely in-memory or affected by the cost of accessing data from network, memory or disks (time-series algorithms usually operate on streaming data and are bound by network latencies). For these exemplars, the amount of computation usually increases (in proportion based on the algorithmic complexity) as the amount of data is increased. Thus, in such cases, *bigger* data translates into *bigger* compute as well. The remaining exemplars are memory-bound when the data is in-memory and I/O-bound when data is on disks. Thus, accelerating analytics workloads requires a holistic approach that address these interconnected bottlenecks: memory, interconnect, compute, and storage. Traditional acceleration approaches take one or more of these paths: (1) Algorithmic modifications to exploit the available hardware resources; (2) Improving performance of existing functions using better hardware; and (3) Employ new algorithms that can exploit novel architectures.

Approaches in the first path involve techniques to parallelize the existing algorithms to enable them to exploit multiple computing resources (e.g., multiple cores on a processor or multiple

CPU nodes). The type of parallelization approach depends on the computation and runtime properties of the exemplars. A majority of analytics exemplars use shared state to represent either the program state or a cost function to be optimized. Among the data structures used for implementing the shared state, only matrices are amenable for scalable distributed-memory parallelization. A majority of analytics exemplars operate on sparse data, materialized in memory using either sparse matrices or specialized data structures such as prefix trees (Lakshmanan et al. [2003]). Operations involving sparse data structures involve accesses via indirection arrays that can generate non-contiguous memory accesses. A number of algorithms that use sparse data structures, in particular those that operate on graphs, generate gather-scatter memory access patterns. These issues limit the use of distributed-memory parallelization approach to the exemplars that use matrices to represent shared state (e.g., Regression, Clustering, OLAP, etc.) Among all examplars, only the Monte Carlo methods are inherently embarrassingly parallel, while some approaches need special reformulation to eliminate execution dependencies, e.g., using the Alternate-Direction Multiplier Method (ADMM) to parallelize optimization problems (Boyd et al. [2011]). In those scenarios, the exemplars are operating on out-of-core data sets, techniques such as MapReduce (Bekkerman et al. [2011], Leskovec et al. [2014]) can be exploited. However, the MapReduce approach is ideally suited for non-iterative, embarrrassingly-parallel algorithms that run in the batch mode.

The second approach involves using better hardware to accelerate key performance components of the exemplars. For example, one could use solid-state drives (SSDs) for accelerating read-only I/O accesses. The SSDs would also improve performance of non-contiguous disk accesses that may be generated during computations involving sparse data. The exemplars that are memory-bound would benefit from improved memory hierarchies: deeper cache hierarchies and larger main memory systems. Finally, time-series processing would benefit from faster networking systems with such as InfiniBand that allow remote direct-memory access (RDMA).

Finally, techniques in the third approach involve using algorithmic techniques that exploit dedicated hardware accelerators, such as single-instruction multiple-data (SIMD) instructions (e.g., x86 AVX or Power VSX), GPUs, FPGAs, and ASICs. Unlike the first two approaches, this approach is more suited toward accelerating key computational kernels and functions. As illustrated in Table 3.2, the analytics exemplars exhibit several repeated computational kernels and functions that can be accelerated using hardware accelerators. Examples of such kernels include various matrix operations such as matrix-matrix, matrix-vector multiplications, linear solvers, factorization. These can be easily accelerated using accelerated libraries such as Intel MKL, ESSL, or CUBLAS which use data-parallel features of SIMD or GPUs. Other kernels that can be accelerated using SIMD or GPUs include FFT, convolution, hashing, sorting, and random number generators. GPUs and SIMD capabilities can be also used for accelerating various functions used by the exemplars, e.g., various distance functions, set operations, aggregation and statistical functions (e.g., MIN, MAX, Average), and bit-vector operations. Several of these functions can be also accelerated by specialized implementations on FPGAs or ASICs. In particular, FPGAs are

Table 3.4: Opportunities for parallelizing and accelerating analytics exemplars.

Model Exemplar	Bottleneck	Acceleration Requirements and Opportunities
Regression Analysis Clustering Nearest-Neighbor Search Recommender Systems Neural Networks Support Vector Machines	Compute-bound I/O-bound	Shared- and distributed-memory task parallelism Data parallelism via SIMD or GPUs Faster I/O using solid state drives
Association Rule Mining	Memory-bound I/O-bound	Shared-memory task parallelism Faster I/O using solid state drives Larger, deeper, and faster memory hierarchies Faster bit operations or tree traversals via FPGAs
Decision Tree Learning	Memory-bound I/O-bound	Larger, deeper, and faster memory hierarchies
Time Series Processing	Compute-bound Memory-bound	Shared-memory task parallelism Data parallelism via SIMD or GPUs High-bandwidth, low-latency memory and networking Pattern matching via FPGA
Text Analytics	Compute-bound Memory-bound I/O-bound	Shared- and distributed-memory task parallelism Data parallelism via SIMD or GPUs Larger, deeper, and faster memory hierarchies Faster I/O via solid state drives Pattern matching and string processing via FPGA
Monte Carlo Methods	Compute-bound	Shared- and distributed-memory task parallelism Data parallelism via SIMD or GPUs Faster bit manipulations using FPGAs or ASICs
Mathematical Programming	Compute-bound	Shared-memory task parallelism Data parallelism via SIMD or GPUs Larger and deeper memory hierarchies Search-tree traversals via FPGAs
On-line Analytical Processing	Memory-bound I/O-bound	Shared- and distributed-memory task parallelism Data parallelism via SIMD or GPUs Larger and deeper memory hierarchies Pattern matching via FPGAs, Faster I/O using solid state drives
Graph Analytics	Memory-bound I/O-bound	Shared-memory task parallelism Larger and deeper memory hierarchies Massive data-parallelism via GPUs

ideally suited for accelerating computational patterns that involve extensive branching (e.g., those based on finite state machine automata), bit-vector manipulations, or dataflow execution. For example, FPGAs can be used to accelerate string pattern matching functions (useful in text analytics and time series processing), bit manipulation, and tree traversals functions.

These three approaches can be used for building re-usable software components that can be specialized to take advantage of available hardware features. Given the functional flow of the analytics workloads (Figure 1.2), one can build an analytics workload using libraries that employ different analytics algorithms based on user or runtime constraints, where an algorithm can have multiple system implementations, e.g., using shared- or distributed-memory parallelism or using MapReduce. Each implementation can, in-turn, use specialized kernels or functions that can exploit various hardware accelerations such as SIMD, GPUs, or FPGAs. The overall execution can be further improved by using hardware components (e.g., SSDs) suited for individual algorithm. Such hardware-software co-design would then enable optimized analytics solutions that can balance customization and commoditization.

CHAPTER 4

Accelerating Analytics in Practice: Case Studies

4.1 TEXT ANALYTICS

Broadly, text analytics workloads can be classified into two groups based on their computational patterns, namely, those workloads that operate on the text data in native form, and those that operate on the data structures that are derived from the input data (e.g., TF/IDF matrices). These factors determine the type of acceleration employed in a text analytics workload.

Key workloads that use native text processing include natural language processing (NLP), network intrusion detection systems, bio-informatics, and semi-structured text processing (e.g., XML and RDF data). One of the basic tasks of any NLP system is to parse the natural language input. The most common approach for natural language parsing uses syntactic parsing that analyzes the grammatical structure of sentences and predict their parse trees. This approach uses the *Cocke-Kasami-Younger* (CKY) dynamic programming algorithm to identify most likely parse trees for context-free languages. The CKY algorithm uses a two-dimensional array called CKY table to store all possible derivations from the context-free grammar. Computations on the CKY table are highly parallelizable and can also be viewed as matrix multiplication operations (Thompson [1994]). The CKY parser has been parallelized over traditional parallel systems using OpenMP and MPI (Johnson [2011]), FPGAs (Bordim et al. [2003]), and GPUs (Yi et al. [2011]). Currently, the highest performing CKY parser for context-free languages uses GPUs and provides 2 to 5 orders of magnitude improvement over CPU implementations (Canny et al. [2013]).

Another interesting application of text analytics is for network intrusion detection system such as Snort (Roesch [1999]). Intrusion detection systems (IDS) identity malicious incoming traffic via inspecting packet payload for attack signatures. Most IDSs including Snort encode the attack signatures as strings and compare input network packet headers against multiple attack signatures to identify any malicious pattern. In practice, the string matching operation account for up to 70% of Snort execution time. Broadly, string matching algorithms can be classified into two groups based on the underlying data structures: algorithms that construct finite state machines (FSM) via building tree representations (e.g., the Aho-Corasick algorithm builds a prefix-tree (trie) with additional links between various internal nodes), and algorithms that view strings as character arrays and operate on them using set operations, (e.g., Boyer-Moore or Boyer-Moore-Horspool algorithms). FSM-based algorithms are more amenable for hardware implementations using FPGAs, whereas the set-oriented string matching algorithms are more amenable to acceleration via SIMD. In particular, both Intel AVX and Power VSX SIMD instruction sets have been used to accelerate string matching problems (Ladra et al. [2012]) . Over the years, there have also been several efforts to accelerate FSM-based string and regular expression matching functions in intrusion-detection systems using FPGAs (Aldwairi et al. [2005]). A recent activity in this space is the emergence of specialized processors designed to accelerate non-deterministic finite automata problems, e.g., the Micron Automata processor (Dlugosch et al. [2014]). This processor has been shown to accelerate a number of finite-automata based text analytics problems (Roy and Aluru [2014]. In general, FSM-based algorithms are also applicable to the general problem of regular-expression matching for text analytics. Regular expression matching has several uses including intrusion detection and bio-informatics (Atasu et al. [2013]). Another interesting use of FSM-algorithms is for processing XML documents. The XML execution model views an XML document as a rooted ordered tree. Various XML query languages such as XPath and XQuery then navigate the XML document tree. Both parsing input XML documents and XML tree traversals use finite-state automata. Both of these operations are amenable to FPGA acceleration and there are several systems in use (e.g., the DataPower XML accelerator) that exploit FPGA to specifically accelerate XML computations. FPGA acceleration of XML processing is particularly attractive in streaming domain, e.g., in the web-services scenario, where incoming XML packets need to parsed and processed in real-time (Dai et al. [2010]).

The second class of text analytics workloads, e.g., text classification, clustering, semantic and topic analysis, uses matrices to summarize key information about the underlying text document corpus and operates on these matrices to get relevant information (Berry et al. [1995]). Key matrix-based operations for text analytics include Singular Value Decomposition, Non-negative Matrix Factorization, and Eignenvector computations (e.g., PageRank). In most cases, the matrices are sparse, and performance of these operations on traditional multi-core CPUs is very poor. In these cases, GPUs have been shown be very effective for improving the performance of matrix computations (Zhang et al. [2009]).

Further Reading
Algorithms: Parsing (Canny et al. [2013], Thompson [1994]), Snort (Roesch [1999]), Matrix Algorithms (Berry et al. [1995]); **Accelerator Systems:** SIMD exploitation (Ladra et al. [2012], Salapura et al. [2012], Shi et al. [2011]), FPGA (Aldwairi et al. [2005], Atasu et al. [2013], Bordim et al. [2003], Court and Herbordt [2007], Cronin [2014], Dai et al. [2010], Mitra et al. [2009a], Roy and Aluru [2014], Schlegel et al. [2013]), GPUs (Canny et al. [2013], Johnson [2011], Kysenko et al. [2012], Yi et al. [2011], Zhang et al. [2009]), Micron Automata (Dlugosch et al. [2014], Roy and Aluru [2014]).

4.2   DEEP LEARNING

Deep learning is perhaps the most exciting and practical area of computer science right now. Broadly, deep learning refers to an architecture which is built using multiple layers of machine learning components, e.g., neural networks with many hidden layers (Bengio [2009]). A deep learning system is designed to learn complex functions that represent high-level abstractions such as images, speech, or languages. In practice, a deep learning system is built using different types of neural networks, e.g., convolution neural networks and fully-connected perceptrons (FLPs), and can be very deep (e.g., GoogLeNet has 22 layers; Szegedy et al. [2014]). In this section, we discuss various approaches in accelerating deep learning systems.

Most commercial installations of deep learning systems are geared toward addressing problems in the consumer space such as intelligent image and video processing (e.g., the Clarifai image processing system) and speech recognition systems e.g., Deep Speech by Baidu, IBM Watson Speech Recognition, or Google Now (Hinton et al. [2012], Ng [2015]). The basic goal of these systems is to accurately classify input query objects (e.g., identify the breed of dog from a set of dog pictures or identify phonemes from a set of audio samples). Some systems are also capable of providing advanced features such as finding similar images or speech-to-text transcription. To accurately classify an object, the system needs to first extract as many of its features as possible. Therefore, most deep learning systems have an hybrid architecture, where the initial stages perform feature extraction and the later stages perform classification. The feature extraction layers either use the convolution or recurrent neural networks. The classification layers are implemented using fully-connected perceptrons (FLPs). Figure 4.2 represents a typical speech processing deep learning system which has two convolution layers and four fully connected layers. The system takes audio speech signal as an input and identifies the corresponding phoneme. First, the speech signal gets pre-processed into samples of frequency spectrogram which captures the key coarse-grained features of the input signal. These samples are then processed by two layers of convolution neural networks that capture fine-grain features of the input data. Output of the convolution layers is then classified by a series of FLPs. The output of this system is a one of the pre-determined phoneme classes (a typical number of classes is 32 K).

The deep learning systems use supervised training approach to train the neural network models (a network model broadly refers to the matrices used to represent the state of the system) Training neural network usually involves solving a non-convex cost function. Most systems solve this optimization problem using variants of the gradient descent algorithm, e.g., the stochastic gradient descent algorithm. The training process involves a forward pass to compute the current state of the system, and a backward pass that updates the weights of the system based on the gradient descent solution. Computationally, both forward and backward passes involve matrix-matrix and matrix-vector multiplications over large single-precision matrices (convolution operation can be implemented using FFT as well; Vasilache et al. [2014]). In practice, to improve the classification accuracy, deep learning systems use large training datasets, have deeper network layers, and have large models. Further, certain neural network models, e.g., convolution neural networks, are

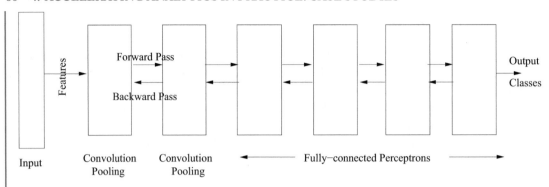

Figure 4.1: Architecture of a deep learning system.

far more computationally intensive than the traditional fully-connected perceptrons. For example, in a speech processing deep learning system, around 30% of the end-to-end time is consumed by the convolution neural network stages. Each convolution neural network layer requires $3 * 10^9$ floating point operations per a minibatch of training data per iteration. Each iteration operates over multiple hours of training data, where each hour corresponds to around 1500 minibatches, and it takes around 30 iterations to converge (van den Berg et al. [2015]). Thus, the total amount of computation required for training a deep learning pipeline is extremely large.

Thus, training a deep learning system is a big data problem in which larger datasets lead to even bigger computational requirements due to the computationally expensive kernels. Clearly, the only way of improving the performance is to parallelize the overall computation and accelerate individual kernels. There are three common approaches for parallelizing the training process: (1) model parallelism in which the model (i.e., the matrices) are partitioned and individual sub-matrices are then trained over the entire training set; (2) data parallelism in which the entire model is trained concurrently using distinct pieces of the training set (called minibatches); and (3) hybrid parallelism that uses both model and data parallelism (Krizhevsky [2014]). A deep learning system is usually built as a cluster of CPU or GPU nodes. Although one can build a deep learning system as a cluster of CPUs (e.g., Project Adam (Chilimbi et al. [2014]) or using the Blue Gene supercomputer (Chung et al. [2014])), a hybrid system with CPUs and GPUs is more suitable. Given the heavy single-precision computational training workload, GPUs are primarily used for accelerating the core computational kernels such as convolutions and matrix multiplications (Chetlur et al. [2014], van den Berg et al. [2015]). Exploitation of GPUs for deep learning is further enhanced as all the key deep learning programming frameworks (e.g., Theano, Caffe, and Torch) provide support for GPU acceleration. Given the high single-precision capabilities of GPUs, one can train the same sized deep learning system using a smaller cluster. For example, the 16 thousand node CPU cluster used by the Google Brain system is now replaced by a much smaller cluster of GPU and CPU nodes (Dean [2015]). Similar design approach is followed by other commercial deep learning systems, e.g., the Baidu Minwa machine is built as a cluster of 36 nodes, each with 2 6-core Intel Xeon E5-2620 processors, 4 Nvidia K40 GPUs, connected by a high-performance low-latency InfiniBand network (Wu et al. [2015]).

Although GPUs are ideal for accelerating deep learning workloads, they suffer from high power consumption. The power consumption is very important while designing datacenter-scale deep learning systems. In such scenarios, accelerating deep learning workloads via low-power FPGAs or specialized ASICs becomes more attractive. Although FPGAs or ASICs cannot match GPUs in absolute performance, they can provide comparable or even better performance per power (a Nvidia K40 GPU consumes around 235W while an FPGA would consume around 25 W). Recent studies (Farabet et al. [2010], Ouyang et al. [2014], Ovtcharov et al. [2015], Zhang et al. [2015]) have demonstrated efficacy of using FPGAs for accelerating convolution neural network computations. The FPGA implementation enables (1) support for multiple layer configurations at run-time, (2) on-chip communication network that minimizes memory traffic to off-chip memory, and (3) spatially distributed array of processing elements that can scale up to thousands of units. In a typical FPGA implementation, input values are first loaded into a multi-banked input buffer. These values are then streamed to the processing elements which then independently perform the matrix computations in the convolution step. The results are then accumulated and re-routed via a specialized network-on-chip back to the input buffers for the next round of computations (Ovtcharov et al. [2015]).

Another area of active research is the development of specialized accelerators for neural network workloads. Examples of such accelerators include DaDianNao (Chen et al. [2014b]), NeuFlow (Pham et al. [2012]), and IBM TrueNorth (Seo et al. [2011]). Both DaDianNao and NeuFlow implement the dataflow versions of different stages of convolution neural networks (e.g., convolution and pooling) in hardware. The IBM TrueNorth, on the other hand, implements a network of integrate-and-fire spiking neurons that have binary outputs. The chip has 4096 processors, each of which has 256 integrate-and-fire spiking neurons (with binary output) each with 256 inputs. Each processor can compute all 256 neurons 1000 times per second asynchronously, with the power consumption of around 100 mW. Unlike DaDianNao and NeuFlow, TrueNorth is designed only for classification, not training.

Further Reading
Basic ideas: Bengio [2009], Schmidhuber [2014]; **Software Infrastructures:** cuDNN (Chetlur et al. [2014]), Theano (Bastien et al. [2012], Bergstra et al. [2010]), Tourch7 (Collobert et al. [2011]), and Caffe (Jia et al. [2014]); **Commercial Solutions:** Google Brain (Le et al. [2012]), Google LeNet (Szegedy et al. [2014]), Clarifai (Zeiler), Deep Speech, Google Now and IBM Watson Speech Recognition (Dean [2015], Hannun et al. [2014], Hinton et al. [2012], Ng [2015], van den Berg et al. [2015]); **Systems:** Google Brain (Dean [2015], Ng [2015]), Project Adam (Chilimbi et al. [2014]), and Baidu Minwa (Wu et al. [2015]);

Specialized Hardware: FPGA (Farabet et al. [2010], Ouyang et al. [2014], Ovtcharov et al. [2015], Zhang et al. [2015], DaDianNao (Chen et al. [2012, 2014a,b], Liu et al. [2015]), NeuFlow (Pham et al. [2012]), TrueNorth (Merolla et al. [2011], Seo et al. [2011]).

4.3   COMPUTATIONAL FINANCE

Computational (or quantitative) finance broadly covers computational techniques that address key problems in financial and insurance domains. These problems can be classified into (1) Simulation, (2) Estimation, (3) Valuation, (4) Calibration, (5) Asset liability management, and (6) Risk Management (Forsyth [2014], Korn [2014]). Specific examples of these problems include simulation (pricing) of various financial instruments (e.g., stock prices, options, interest rates, commodity prices); valuation of derivatives; quantitative methods for high-frequency algorithmic trading; measuring the risk of a whole bank or an instrument, and managing asset liabilities for life insurers or pension funds.

One of the most common computation patterns in computational finance is the use of stochastic differential equations (SDEs) for pricing. The most widely used model, the Black-Scholes model, models the price variations (paths) of a stock over the time as a geometric Brownian motion model using market parameters such as interest rates, drift, and volatility. Unfortunately, the Black-Scholes model assume constant volatility and hence, does not provide an accurate reflection of modern financial markets. The Heston Model extends the Black-Scholes model by using a second stochastic differential equation to capture stochastic volatility variations (Delivirias [2012]). Solving these differential equations via finite difference methods is computationally very expensive, and in many cases, such as pricing exotic options, it is not possible to compute a closed form solution. In such cases, approximate approach using Monte Carlo (MC) simulation is employed. In the MC approach, the first step generates a set of random price paths, the second set calculates the associated value for each path using a payoff function, the third step averages these values, and discounted to a given day to compute the option value for that day. The MC approach usually requires a pseudo-/quasi-random number generator to generate data with a specific distribution (e.g., normal). Both SDE solving and MC approaches involve significant floating point computations and are very computationally expensive.

The choice of accelerators for computational finance workloads depends not only on the computational complexities of the kernels, but also on the runtime and operational constraints. Running these workloads on servers or clusters of standard multi-core CPUs is not feasible either due to long running times or energy consumption (Christian de Schryver and Schmidt [2011], Wehn [2011]). Earlier, many of the modeling tasks were done in batch mode. But now, many of these models have to be executed in real-time. Further, many of the workloads, e.g., for algorithmic trading, require ultra-low latency, wire-speed end-to-end performance. In addition, there are additional reporting requirements that need to met to satisfy various regulations. Given the high computational requirements, computational finance workloads are ideal to be accelerated using traditional compute-accelerators such as GPUs and Intel Xeon-Phi (Giles [2010], xcelerit [2014]). However, given the high power consumption of these accelerators, FPGAs - with their low-power footprint, are proving to be a popular choice for accelerating computational finance workloads (Luk [2013]).

The most common analytical model in computational finance is the Monte Carlo simulation. The Monte Carlo approach is inherently embarrassingly parallel and can be easily parallelized. However, Monte Carlo methods require high-quality cheap source of pseudo-random numbers, particularly for key probability distributions: uniform, Gaussian, exponential, and lognormal (Thomas et al. [2009]). In some cases, the Monte Carlo method needs low-disperency quasi-random number generators such as Sobol sequences (Bordawekar and Beece [2015]). The computational characteristics of these random-number generators depend on many factors such as period, statistical quality and computational costs. Ideally, one would like to have a random number generator with high statistical quality and period, but with low computational costs. An approach that achieves such balance is the binary linear generator that performs binary linear operations (logical conjunction or exclusive disjunction) on vectors of individual bits. Such operations can be implemented using bit-wise operations such as masking, exclusive-or, and shifting. Generation of non-uniform random number generators use different approaches, e.g., inversion, transformation, rejection, and recursive (Thomas et al. [2009]). The first three generate non-uniform random numbers by consuming uniform input and operating on it, e.g., inverting via applying inverted Cumulative Distribution Function (CDF) to a uniform sample, transforming a fixed set of uniform samples into a fixes set of non-uniform samples using transcendental functions (Box-Muller transform), select a only few values from a large set of uniform numbers (the Ziggurat approach). The recursive approach directly generates non-uniform samples, e.g., the Wallace method can generate Gaussian or exponential distribution efficiently without using any transcendental functions. However, this approach has problems with correlation between output sample values, hence it is not used widely. While both uniform and non-uniform random number generators have shown to be accelerated via CPU, GPU, and FPGAs (Matsumoto and Nishimura [1998], Thomas et al. [2009]), FPGAs have clear advantages due to the availability of very fine-grain binary linear operations (both CPUs and GPUs have word-based instructions). Traditionally, CPUs have been shown to accelerate both uniform and non-uniform random number generators, e.g., the Mersenne Twister uses SIMD instructions to achieve a higher generation rate. While GPUs can effectively accelerate uniform random number generators, they suffer from the cost of branching inherent in certain non-uniform random number generators, e.g., the Ziggurat approach.

Given the ability of FPGAs to effectively generate high-quality random numbers while consuming low power and their affinity to process data in low-latency scenarios, FPGAs have the accelerator of choice in the financial industry. In practice, FPGAs have been used to accelerate option and derivative pricing (xcelerit [2013b]), high-frequency trading (Leber et al. [2011]), Collaterized Debt Obligations (CDOs) pricing (Kaganov et al. [2011]), real-time risk management (Studnitzer and Mencer [2013]), and Credit Value Adjustment (CVA) computations (Kaganov et al. [2011]). Recently, GPUs (Giles [2010]) and other accelerators such as Intel Xeon Phi (xcelerit [2014]) have been increasingly used in the financial modeling applications. Although the raw compute performance of such compute accelerators is better than FPGAs, power consumption and cost of invoking the accelerator functions still remain as issues for GPUs.

Further Reading
Basic Idea: Christian de Schryver and Schmidt [2011], Forsyth [2014], Korn [2014], Wehn [2011] **Algorithms:** Heston Model (Delivirias [2012]), CVA (Kaganov et al. [2011]), Pseudo-random number generators (Matsumoto and Nishimura [1998], Thomas et al. [2009]), Sobol Sequences (Bordawekar and Beece [2015]), CDO (Kaganov et al. [2011]); **Accelerator Systems:** SIMD exploitation (Matsumoto and Nishimura [1998], FPGA (de Schryver et al. [2011], Jin et al. [2011], Leber et al. [2011], Morris et al. [2009], Sadoghi et al. [2010], Studnitzer and Mencer [2013], Wang et al. [2013], Weston et al. [2010, 2011], xcelerit [2013b]), **GPUs** (Fricker [2015], Giles [2010], Grauer-Gray et al. [2013], Lahlou [2013], Lotze et al. [2012], Papamanousakis et al. [2015], xcelerit [2013a], and Intel Xeon-Phi (xcelerit [2014]).

4.4   OLAP/BUSINESS INTELLIGENCE

The most common scenario of OLAP workload involves processing large amount of disk-resident data stored in relational tables using relational queries written in SQL (ROLAP). A related, but less widely used, strategy involves viewing the disk-resident data as multi-dimensional arrays (MOLAP). With availability of cheap, and reliable DRAM chips, main memory databases are becoming increasingly popular (Lahiri [2014]). While the traditional database performance is affected by disk I/O costs, performance of main-memory and streaming databases is affected by memory and networking costs.

In OLAP workloads, the opportunities for acceleration are dependent on a variety of factors which include (1) type of underlying logical model (i.e., relational, multi-dimensional, or hybrid), (2) execution scenario (e.g., disk-based, in-memory, or streaming), (3) system implementation issues (e.g., row-based or columnar storage), and (4) type of function to be accelerated (e.g., query compilation or execution and various utility functions, e.g., sorting).

In practice, database systems use accelerators for improving both I/O and compute costs. For example, the IBM Netezza system uses a combination of FPGA-based acceleration and customized software to optimize data-intensive mixed database and analytics workloads with concurrent queries from thousands of users. The Netezza system uses two key principles to improve performance: (1) reduce unnecessary data traffic by moving processing closer to the data, and (2) use parallelization techniques to improve the query processing costs (Feldman [2013]). A Netezza appliance is a distributed-memory system with a host server connected to a cluster of independent servers called the snippet blades (S-Blades). A Netezza host first compiles a query using a cost-based query optimizer that uses the data and query statistics, along with disk, processing, and networking costs to generate plans that minimize disk I/O and data movement. The query compiler generates executable code segments, called snippets which are executed in parallel by S-blades. Each S-blade is a self-contained system with multiple multi-core CPUs, FPGAs, gigabytes of memory, and a local disk subsystem. For a snippet, the S-Blade first reads the data from disks into memory using a technique to reduce disk scans. The data streams are then processed by FPGAs at wire speed. In a majority of cases, the FPGAs filter data from the original stream using predicate evaluation, and only a tiny fraction is sent to the S-Blade CPUs for further processing. The CPUs then execute either database operations like sort, join, or aggregation or core mathematical kernels of analytics applications on the filtered data streams. Results from the snippet executions are then combined to compute the final result. The Netezza system can operate on thousands of data streams in parallel. In addition to filtering, Netezza also exploits FPGAs to accelerate key utility functions such as decompression.

When the data is entirely in memory, the overall execution becomes memory-bound. To improve memory performance, databases implement various optimization techniques such as specialized memory layouts such as columnar storage (Manegold et al. [2000]) and operating on compressed data (Raman et al. [2013]). These techniques improve the memory access locality and enable compute acceleration using SIMD intrinsics such as x86 AVX2 and Power VSX. For in-memory databases, SIMD intrinsics have been shown to accelerate various key operations in query execution, e.g., computing set intersections, predicate evaluations, and aggregation calculations (Lahiri [2014], Raman et al. [2013], Schlegel et al. [2013], Sikka et al. [2013], Willhalm et al. [2009b], Zhou and Ross [2002]). SIMD instructions are also applied for accelerating key utilities such as sorting, compression/decompression, and string processing (Inoue et al. [2007]). A special case of main memory databases is streaming database that operates on (potentially infinite) streams of data using a pre-defined set of queries (non-streaming databases support ad-hoc queries as well). Many of these queries perform filtering or matching operations that are very amenable for SIMD exploitation (Gedik et al. [2008], Wang et al. [2010]).

For in-memory MOLAP systems, aggregation over large datasets is often the performance bottleneck. In MOLAP systems, data is viewed logically as a multi-dimensional cube and processed using operators that compute on regions built as a collection of cells. The MOLAP data is usually sparse, and stored using specialized sparse data structures. Aggregating over such sparse datasets often results in accessing non-contiguous (strided) data, which makes efficient exploitation of SIMD intrinsics difficult. GPUs, on the other hand, can use their massive data-parallelism capabilities and high-memory bandwidth to parallel aggregation of strided data. An example of this approach is the Jadox Palo in-memory MOLAP engine (Strohm [2015]) which uses GPUs for aggregation over large strided datasets. Recently, similar GPU techniques have been explored for accelerating OLAP queries on relational data (Wu et al. [2014b]).

Recently, there has been renewed interest in exploring specialized hardware acceleration support for database operations. Key examples of such approaches include Sonoma (Vinaik and Puri [2015]), HASHI (Arnold et al. [2014]), and Q100 (Wu et al. [2014c]). The Oracle SPARC Sonoma processor incorporates a novel Database Accelerator (DAX) that operated on in-memory decompressed and compressed columnar vectors in streaming manner. The DAX architecture also included new SIMD instructions designed to accelerate core database operations. The HASHI approach describes new instruction set instructions for speeding up 32-bit hash functions for integer and string keys. The Q100 proposal describes an architecture of a DataBase Processing Unit (DPU), built as a collection of heterogeneous ASIC tiles that process relational tables quickly and with low power.

Further Reading
Algorithms and Software Solutions: Exploiting SIMD for accelerating database kernels (Schlegel et al. [2013], Willhalm et al. [2009b], Zhou and Ross [2002]); **Systems:** Netezza (Feldman [2013]), Oracle (Lahiri [2014]), SAP Hana (Sikka et al. [2013]), DB2 BLU (Raman et al. [2013]), MonetDB (Manegold et al. [2000], InfoSphere Streams (Gedik et al. [2008]); **Specialized Hardware:** Oracle Sonoma (Hetherington [2015], Vinaik and Puri [2015], Q100 (Wu et al. [2014c]), HASHI(Arnold et al. [2014]).

4.5 GRAPH ANALYTICS

Similar to text analytics, graph analytics workloads can also be partitioned into two classes based on their computational patterns. The first approach navigates the graph explicitly and is used for addressing pattern matching and traversal algorithms. The second approach uses sparse matrix-based linear algebraic solutions for solving structural graph analytics problems. Both approaches result in a large number of high-latency small non-contiguous memory accesses, making graph analytics a memory-bound problem, with very limited spatial and temporal memory localities. A way of accelerating memory-bound problems is to use processors which support massive multi-threading with high memory bandwidth, and thus, can effectively hide or tolerate memory latencies by overlapping computation and memory accesses of multiple threads.

Examples of such processors include the Cray Threadstorm processor and GPUs. The Cray Threadstorm processor is a massively multi-threaded processor that is used to power the Cray XMT system (Kopser and Vollrath [2012]). The Cray XMT is a distributed shared memory system built as a cluster of 4 Thunderstorm processors, connected to a very high-speed interconnect organized as a torus. Each cluster node can have upto 64 GB of memory and the overall system can be scaled to multiple TByte of main memory. From an user's perspective, the XMT appears as a single processor with a large number of threads operating in a shared address space. On the Cray XMT, threads are lightweight software objects that are mapped onto hardware streams.

A stream has its very small register state and executes its instructions independently. Typically, the compiler generates many more threads than the number of streams in the machine that are then multiplexed onto the hardware streams. The small per-thread state allows lightweight context switching at every instruction cycle. The uRiKA graph analytics system is built using the Cray XMT infrastructure (Maltby [2012]). The uRiKA system is designed specifically to accelerate queries on in-memory RDF (Resource Description Framework) databases. RDF is a W3C standard, designed to enable semantic web searching and integration of disparate data sources. The Semantic Web is a graph that captures relationships between entities using triplets (subject, predicate, object) where each triplet is essentially an edge (predicate) connecting two nodes (subject and object). The RDF representation is used extensively to represent knowledge graphs, e.g., biological network graphs. The RDF databases are queried by a specialized query language called SPARQL which enables matching of graph patterns the RDF databases. On the uRiKA system, the SPARQL queries are parallelized by the compiler into multiple small queries which are then executed by the underlying Cray XMT system. The uRiKA system has been used extensively to accelerate semantic web workloads on very large graphs, e.g., large-scale data mining of pharmacological, chemical, and biological semantic graphs (Henschel et al. [2014]).

The Single-Instruction Multiple-Thread (SIMT) execution model of the Nvidia GPUs is very similar to the multi-threaded model supported by the Cray Thunderstorm processor. The current versions of Nvidia GPUs exhibit memory bandwidth over 280 GB/s and can support millions of threads. Over the years, several traversal-based graph analytics algorithms have been ported on the Nvidia GPU, e.g., Breadth-First Search, Single-source Shortest Path, All-pairs Shortest Path, and Minimum-Spanning Tree (Harish and Narayanan [2007], McLaughlin and Bader [2014], Merrill et al. [2012]). There are multiple graph analytics libraries that provide efficient implementations of these and other graph analytics algorithms (e.g., Gunrock and MapGraph; Fu et al. [2014], Wang et al. [2015]). These implementations represent graphs using array-based data structures such as adjacency list, v-graph (vector-graph; Blelloch [1990]), or structure of arrays, and use either the BSP (Bulk Synchronous Programming) or GAS (Gather-Access-Scatter; He et al. [2007]) approaches to navigate the graphs. Although these implementations suffer from lack of spatial or temporal memory localities, they are aided by many of the GPU's architectural features such as massive multi-threading via SIMT, effective thread scheduling, large register files and shared memory, and texture memory/read-only caches. In particular, GPU's texture memory provides hardware support for improving performance of non-contiguous memory accesses. Unfortunately, all the current implementations work on the graph datasets that can fit GPU's device memory. Development of scalable multi-GPU out-of-core graph traversal algorithms is not trivial, and is an area of active research (Wang and Owens [2013]).

The second approach for implementing graph analytics involves operating on sparse matrix representations of the input graphs. The linear algebraic approach is suitable for computing various structural properties of a graph, e.g., the betweenness centrality which uses algorithms to compute eigenvectors of the graph matrix (e.g., PageRank; Bryan and Leise [2006], Mahoney).

The key kernels used by this approach include sparse matrix-vector multiplication (SpMV) and sparse matrix-dense matrix multiplication (`csrmm`). Efficient implementations of these and other related kernels are available both on CPU and GPU platforms (e.g., cuSPARSE, Intel MKL, and IBM ESSL libraries). As sparse matrix computations are also memory bound, performance of these kernels on GPUs is significantly better than on current generation of multi-core CPUs (Yang et al. [2011]). Recently, a specialized graph processor architecture (Song et al. [2013]) was proposed to address the key weaknesses in the traditional CPU designs. The proposed graph processor is a specialized parallel processor that uses a new instruction optimized for sparse matrix operations. The graph processor is built as an array of sparse matrix processors (called node processors) connected via high-bandwidth three-dimensional communication fabric. Large sparse matrices are distributed over these node processors and operated using the new instruction set. Each node processor has cache-less local memory. All data computations, indices-related computations, and memory operations are handled by specialized accelerator modules rather than by the central processing unit. The processors use new message-routing algorithms that are optimized for communicating very small packets of data such as sparse matrix elements or partial products. While current performance estimates are based on simulation, the specialized graph processor direction is very promising and deserves further investigation.

Further Reading
Algorithms and Software Solutions: Graph algorithms on GPUs (Harish and Narayanan [2007], Harish et al. [2009], McLaughlin and Bader [2014], Merrill et al. [2012], Yang et al. [2011], Page Rank (Bryan and Leise [2006]), Gunrock (Wang et al. [2015]), MapGraph (Fu et al. [2014]), v-graph (Blelloch [1990]), GAS (He et al. [2007]); **Systems:** Cray XMT (Kopser and Vollrath [2012]), uRIKA (Henschel et al. [2014], Maltby [2012]); **Specialized Hardware:** Song et al. [2013].

CHAPTER 5

Architectural Desiderata for Analytics

In the previous chapters, we analyzed behavior of existing analytics workloads, examined their computational and runtime patterns using analytics exemplars, and discussed various acceleration opportunities. In this chapter, we discuss future trends in analytics workloads and its impact on designing future processors and systems specifically targeted for analytics.

Within the span of a few years, the analytics applications have moved from being tools of *convenience* to being *essential* tools. Advances in systems, software, and hardware have drastically changed the usage and types of analytics workloads, as outlined here.

- Widespread availability of cheap and reliable internet services has enabled the *low-power* mobile devices to become the dominant platform for consuming analytics applications and as well as for generating data for analytics workloads. With the emergence of Internet-of-Things, e.g., home thermostat, and *wearable* devices, this trend is going to continue in foreseeable future. Further, analytics on mobile devices would become more common-place in enterprise scenarios (e.g., point-of-sale devices).

- Data being consumed or generated by the analytics workloads is accessed primarily from virtualized *cloud* resources. Cloud enables scalable user access to ever-increasing data repositories. While the cloud infrastructure has enabled easy sharing of data for analytics workloads, it has created serious issues with data privacy and security.

- Advances in software-hardware co-design has enabled complex resource-intensive *multi-modal* applications to become ubiquitous, e.g., *Intelligent Personal Assistants* (IPAs) that use inputs such as voice, vision, and contextual (e.g., spatial coordinates) information to provide answers in natural languages (Hauswald et al. [2015]). As a consequence, it has opened up new domains for exploiting analytics, e.g., personalized healthcare, or home and car automation.

- Convergence of *mobile* and *social* domains has lead to increasing use of spatial and temporal analytics workloads that often require executing analytics queries in (near) real-time. For example, one can imagine a trip scheduler that takes voice input and generates personalized travel itinerary in real-time based on specified constraints (i.e., price, time, etc.) and historical travel data.

- Analytics workloads are increasing using operations that overlap the traditional analytics, high-performance computing, and data management boundaries. For example, current state-of-the-art techniques for parallelizing the gradient descent algorithm borrow heavily from parallel computing and distributed data management approaches (Niu et al. [2011]).

- End-to-end workflow of analytics workloads exhibit components with different runtime constraints: a front-end component executing on low-power devices generating streams of requests to the back-end components processes these requests either purely in-memory or in out-of-core manner. Some of the components may be executing on a shared memory system, some on a distributed cluster or some may be executing on a cloud environment. Thus, the end-to-end workload would use multiple acceleration strategies with different characteristics.

5.1   ACCELERATORS FOR ANALYTICS WORKLOADS

To address the requirements discussed in the previous chapter, it is important to review currently available accelerator options and if they can satisfy requirements of current and future analytics workloads.

Broadly, analytics accelerators can be viewed as a hierarchy built using a set of *foundational* accelerators. Foundational accelerators are designed to accelerate core execution functions, such as, compute, memory, I/O, and networking. Examples of foundational accelerators include, compute accelerators such as SIMD engines, GPUs, and FPGAs; memory accelerators such as GPU texture memory or Micron's Active memory (Kirsch [2003]); networking accelerators such as network processors or RDMA-based accelerators (Lu et al. [2014]); and storage accelerators such as active storage using non-volatile memory devices (Fitch [2013]). The foundational accelerators can be used to build higher-order accelerators, namely functional, data-structure, kernel, and workload accelarators (Table 5.1). These accelerators can also be characterized by their location and execution patterns. The accelerators can be co-located on the same die as the host processor (e.g., SIMD) or can be connected to the processor via an external interface such as PCI-E or connected directly to the memory subsystem (e.g., Micron's Yukon; Kirsch [2003]), or connected directly on the network (e.g., using network processors), or connected to storage sub-systems (e.g., BlueGene Active Storage; Fitch [2013]).

- **Functional Accelerators:** Functional accelerator accelerate specific operations or functions. As observed in Table 3.2, the analytics exemplars have several common functions that can be accelerated. For example, regular expression evaluation for pattern matching (Tutomu Murase and Kuriyama [2000]), compression/de-compression, encryption-decryption, etc. Table 5.1 presents a list of functional accelerators and how they can be implemented. While a number of functional accelerators can be implemented via foundational accelerators (e.g., support for cryptography in Intel's AVX2 instruction set). However, functions such as dis-

Table 5.1: Classification of analytics accelerators

Accelerator	Type	Implementation	Location
Pattern/Regular Expression Matching	Functional	CPU, GPU	On/Off-chip, On-Network
Compress and Decompress		FPGA	Near-storage, Near-memory
Encryption and Decryption			
Bit-vector Processing	Functional	CPU, FPGA	On/Off-chip
Distance Metric	Functional	CPU, GPU	On/Off-chip, Near-memory
Streaming Processing	Functional	CPU, FPGA	On/Off-chip, On network
Kernel Functions	Functional	CPU, GPU	On/Off-chip, Near-memory
Hash Tables	Data-structure	CPU, GPU, FPGA	Near-memory, Near-storage
Bloom Filters			
K-d, R, Binary			
Prefix/Suffix Trees			
Key-value Pairs			
Index (B-Tree, Inverse)			
Dense and Sparse Matrices			
FFT and Convolution	Kernel	GPU, FPGA	On/Off-chip Near-memory, Near-storage
Sorting	Kernel	CPU, GPU, FPGA	Near-memory, Near-storage
Matrix Computations	Kernel	CPU, GPU	Near-memory, Near-storage
Near-neighbor Search			
Random Number	Kernel	CPU, GPU, FPGA	On/Off-chip
Top-K Processing	Kernel	CPU, GPU	Near-memory, Near-storage On network
Visualization	Workload	GPU	On/Off-chip, Near-memory Near-storage, On network
Graph Traversal	Workload	GPU, FPGA	Near-memory, Near-storage
Security	Workload	FPGA	Near-memory, Near-storage
OLAP	Workload	CPU, GPU, FPGA	Near-memory, Near-storage
Neural Networks	Workload	GPU	Near-memory
Financial	Workload	GPU, FPGA	On/Off-chip, On network
Bio-informatics	Workload	GPU, FPGA	Near-memory, Near-storage

tance computations, e.g., root-mean square error between two vectors; kernel functions such as sigmod functions, are not supported in hardware.

- **Data-structure Accelerators:** Foundational accelerators can be also used to improve performance of operations on key data structures such as hash-tables, dense and sparse matrices, bloom filters, and a variety of trees which include B-trees, K-d, and R-trees. Examples of non-matrix data structure accelerators include hash table acceleration on GPUs (Alcantara [2011]), bloom filter acceleration on FPGAs (Dharmapurikar et al. [2004]), and K-d trees on GPUs (Foley and Sugerman [2005]). There is excellent support for accelerating various computations on dense and sparse matrices on GPUs: sparse matrix computations are memory bound and can make extensive use of texture memory support on GPUs. Recent studies have investigated alternative strategies for accelerating sparse matrix accesses, e.g., using FPGAs (Fowers et al. [2014]) or via specialized architecture (Song et al. [2013]).

- **Kernel Accelerators:** The functional and data-structure accelerators can be used to build accelerate individual kernels. Examples of kernels that can be accelerated include: (1) sorting, whose comparator and exchange functions can be accelerated by functional accelerator, e.g., using SIMD instructions or via GPU (Inoue et al. [2007], Merrill and Grimshaw [2011]), (2) various matrix computations, e.g., BLAS and sparse matrix functions, (3) FFT and convolution, (4) random number generators, and (4) analytical kernels such as k-Means clustering. Most current CPUs and GPUs support highly tuned libraries to support many of these kernels (e.g., Intel MKL, IBM ESSL, and Nvidia's cuBLAS, cuSPARSE, and cuRAND libraries).

- **Workload Accelerators:** Final type of accelerator accelerates execution of workloads using functional, data-structure, and kernel accelerators. Examples of workload accelerators include visualization, XML processing, security, financial, neural networks, OLAP, and bioinformatics. Such workload accelerators can be built either using FPGAs or GPUs (Putnam et al. [2014], Wu et al. [2014a]). As we discussed in Chapter 4, there has also been significant interest in developing specialized hardware implementations of key workloads, e.g., neural network-based computations, e.g., NeuFlow and TrueNorth processors (Pham et al. [2012], Seo et al. [2011]), OLAP (Wu et al. [2014c]), and XML processing (Mitra et al. [2009b]).

It is clear that while existing processor architectures and systems are able to execute a few analytics workloads, they still lack capabilities to effectively accelerate certain key analytics computational patterns (e.g., sparse matrix computations) and to support additional requirements of the upcoming analytics workloads. Specifically we have the following.

- **Supporting efficient non-contiguous memory accesses:** The most common computational pattern observed across multiple analytics workloads is the non-contiguous memory access pattern caused by computations on sparse matrices, high-dimensional sparse data, or graphs. Current CPU memories assume a linearized layout which leads to inefficient memory behavior. Specialized memory designs such as texture memory supported by GPUs provide two-dimensional memory accesses, but can not sufficiently address random access requirements of graph analysis workloads. Recently, there has been some interest in designing processors specifically for graph analytis workloads, e.g., the Graph Processor Architecture from Song et al. [2013]. However, traditional CPU still lack capabilities to efficiently support non-contiguous memory accesses.

- **Supporting big data computations:** In practice, analytics workloads operate on a wide variety of large datasets, both persistent and streaming. As we observed, a majority of analytics workloads extract useful information from the input data, and compute results that are much smaller than the input (exception being OLAP and association rule mining). The amount of computation is at least O(N) complexity, which means that big data often translates to big compute. Traditional data-intensive algorithms were designed to minimize the number of I/O accesses, but not to optimize the amount of data to be moved. However, with Peta- and Exa-byte data sizes, the cost of moving data to compute cores has become dominant. A way to address this problem is to bring compute functions closer to the data sources (e.g., disks, network streams). Such *active* computations cannot only reduce number of I/O accesses, but can also reduce the total amount data moved (Acharya et al. [1998], Riedel et al. [1998], Uysal et al. [2000]). Recent studies (Fitch [2013], Kirsch [2003]) have explored opportunities for active storage in the context of DRAMs and non-volatile memories. However, far more work is required for building a usable ecosystem for active computing.

- **Supporting approximate computations:** Approximate computing is an emerging paradigm that enables building efficient hardware and software implementations by exploiting implicit resilience of applications toward in-exactness of their underlying computations (Chippa et al. [2013], Nair [2014]). As we have observed in the previous sections, for a wide variety of analytics models, specifically in data mining and machine learning domains, e.g., neural networks, nearest-neighbor search, clustering, support vector machines, etc., *relative* properties of intermediate or final computations are more important than *absolute* values. In such cases, computations done in an approximate manner or using lower-precision arithmatic—as long as they preserve the relative ordering—have little or no impact on the output results (Gupta and Gopalakrishnan [2014], Gupta et al. [2015], Mishra et al. [2014]). Approximate implementations of algorithms can reduce the amount of computation and improve memory performance by reducing memory footprint and improving memory utilization. These improvements can lead to significant performance improvements and lower energy consumption.

- **Support for *multi-tenant* execution:** A consequence of executing in a cloud environment is that the underlying system (computing, networking, and I/O) resources are used for multiple types of workloads (e.g., analytics, data management, and high performance computing). To effectively serve these different workloads simultaneously, the underlying system resources need to shared fairly. For analytics workloads, where accelerators are being increasing used for improving performance of specific kernels, virtualization brings additional challenges. At present, support for virtualizing acceleration resources is still primitive.

5.2   BRINGING IT ALL TOGETHER: BUILDING AN ANALYTICS SYSTEM

So far we have discussed computational and runtime characteristics of different analytics workloads, and presented different strategies for accelerating these workloads. The key questions that is still unanswered is how one can integrate these acceleration solutions into a single analytics system.

As we have observed so far, a majority of analytics workloads exhibit characteristics similar to classical high-performance computing (HPC) workloads and traditional data management systems. Figure 5.2 presents a view of how models from these three domains interact with each other. Both HPC and analytics workloads use mathematical formulations to solve the problem at hand (specifically, both approach make extensive use of linear algebraic kernels.) Some of the analytics models, e.g., modeling, and graph analytics, are widely used in the HPC context. Further, HPC software infrastructure can be used for implementing scalable analytics algorithms, e.g., using MPI as a communication layer. However, analytics and HPC workload differ in a key aspect: a majority of HPC applications have a single domain-specific focus (e.g., seismic processing, cosmological simulations, or computational fluid dynamics). Such HPC applications can be viewed as *information extraction* processes that have a single workflow designed to address domain-specific functional and runtime goals. The application domain determines the runtime characteristics, e.g., input/output data representations, data layout, and sizes. HPC applications also exhibit compute-intensive behavior for in-memory data. Unlike HPC applications, most analytics applications have multi-domain focus and support several independent workflows with potentially different domain-specific functional and runtime goals. Each workflow could have different computational and I/O characteristics. Thus, an analytics application can be viewed as an *information integration* process. Similar similarities and differences exist between data management and analytics workloads. Both analytics and data management share the on-line analytical processing (OLAP) model; unstructured and semi-structured data processing (e.g., XML, RDF, or natural text) share algorithms with text and graph analytics models; and data management over streaming data has similarities with time-series processing. Also, many analytical workloads use data stored in relational databases as the primary source for input. The main difference between data management and analytics workloads is transactional processing that involves management of concurrent update operations. By default, analytics workloads operate on read-only data. Unlike analytics workloads, performance of traditional data management systems is always affected by data access costs (from disks, memory, or network). In addition, transactional management systems often use specialized disk layout and index structures (e.g., B+ trees) that are not needed by analytics workloads. Finally, all three domains share the visualization component, which can be used for viewing input, intermediate data, and results.

Thus, an ideal analytics system should have the following key characteristics:

- a heterogeneous scale-out architecture that supports different types of compute, memory, and I/O devices, and the ability to flexibly choose resources for a given workload;

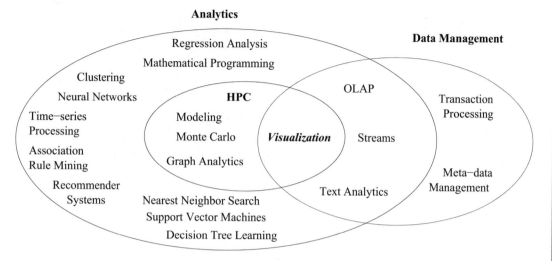

Figure 5.1: Relationships between analytics, high-performance computing, and data management models.

- focus on information integration, not just computational performance. In addition to supporting matrix (sparse and dense) computations, the system should also provide support for data-structures such as hash tables, high-dimensional trees, linked-lists, etc.;

- tighter integration with the data-centric ecosystem, e.g., data warehouses, text repositories, stream processing systems, etc.;

- balanced support for computation, memory, networking, and I/O; and

- should be able to support multiple analytics workloads (multi-tenancy).

Clearly, a traditional scalable HPC system designed for accelerating FLOP-intensive workloads may not be a good analytics system. In contrast, a well-designed analytics system can serve as an HPC system. At present, there are a few scale-out systems that are specialized for individual domains, e.g., Netezza for OLAP (Feldman [2013]) and Google Brain (Le et al. [2012]), Baidu Minwa (Wu et al. [2015]), and Project Adam (Chilimbi et al. [2014]) for deep learning. However, these systems are not flexible enough to provide effective support for analytics workloads. The problem of designing a scalable, flexible, multi-tenant analytics system is open and needs further investigation.

Further Reading

Accelerators: Dharmapurikar et al. [2004], Farroukh et al. [2011], Fitch [2013], Foley and Sugerman [2005], Fowers et al. [2014], Inoue et al. [2007], Jacob and Gokhale [2007], Kirsch [2003], Tutomu Murase and Kuriyama [2000], Lu et al. [2014], Merrill and Grimshaw [2011], Mitra et al. [2009b], Najafi et al. [2013], Putnam et al. [2014], Song et al. [2013], Tumeo et al. [2010], Wu et al. [2014a]; **Approximate Computing:** Chippa et al. [2013], Gupta and Gopalakrishnan [2014], Gupta et al. [2015], Mishra et al. [2014], Nair [2010, 2014]; **Systems:** Google Brain (Dean [2015], Ng [2015]), Project Adam (Chilimbi et al. [2014]), and Baidu Minwa (Wu et al. [2015]); **Specialized Hardware:** FPGA (Farabet et al. [2010], Ouyang et al. [2014], Ovtcharov et al. [2015], Zhang et al. [2015], DaDianNao (Chen et al. [2012, 2014a,b], Liu et al. [2015]), NeuFlow (Pham et al. [2012]), TrueNorth (Merolla et al. [2011], Seo et al. [2011]).

APPENDIX A

Examples of Industrial Sectors and Associated Analytical Solutions

1. Financial Services

   - Core Banking: Customer Insight, Product Recommendation, Fraud Detection and Prevention, Underwriting, KYC, Credit Scoring
   - Payments: Fraud Detection and Prevention, Anti-Money Laundering, Underwriting
   - Financial Markets: Pricing, Risk Analysis, Fraud Detection and Prevention, Portfolio Analysis, Product Recommendation, Merger and Acquisition Analytics
   - Insurance: Risk Analysis, Cause and Effect Analysis, Underwriting, Claims Analysis
   - Financial Reporting: Revenue Prediction, Regulatory/Compliance Reporting, Scorecard/Performance Management

2. Healthcare

   - Drug Interactions, Disease Management, Preliminary Diagnostic Analysis, BioMedical Statistics
   - Healthcare Payer: Insurance Fraud, Clinical Cause and Effect, Medical Record Management, Network Management Analytics
   - Healthcare Provider: Employer Group Analytics, Patient Access Management, Clinical Resource Management, Patient Throughput, Quality and Compliance

3. Retail: Promotions, Inventory Replenishment, Shelf Management, Demand Forecasting, Price and Merchandising Optimizations, Real-estate optimizations, Workforce Efficiency Optimizations

4. Manufacturing: Supply Chain Optimizations, Demand Forecasting, Inventory Replenishment, Warranty Analysis, Product Customization, Product Configuration Management

5. Transportation: Scheduling, Routing, Yield Management, Traffic Congestion Analysis

6. Hospitality: Pricing, Customer Loyalty Analysis, Yield Management, Social-media marketing, Workforce Scheduling

7. Energy: Trading, Supply-demand Forecasting, Compliance, Network Optimizations

8. Communications: Price Plan Optimizations, Customer Retention, Demand Forecasting, Capacity Planning, Network Optimizations, Customer Profitability

9. Integrated Supply Chain: Scheduling, Capacity Planning, Demand-Supply Matching, Location Analysis, Routing,

10. Marketing and Sales Analytics: Customer Segmentation, Co-joint Analysis, Lifetime Value Analysis, Topic/Trend Analysis, Market Experimentation, Yield (Revenue) Analysis

11. Legal Analytics: eDiscovery, Identification, Collection, Record Management

12. Customer Analytics: Customer Retention/Attraction, Pricing Optimizations, Brand Management, Customer Life Cycle Management, Customer-specific Content Specialization, Call Center Voice Analytics

13. Life Sciences: Gene Pool Analysis, Drug Discovery, BioInformatics

14. Human Resources: Churn Analysis, Talent Management, Benefits Analysis, Workforce Placement Optimizations, Call Center Staffing,

15. Government: Fraud Detection, Crime Prevention Management, Revenue Optimizations, Tax Compliance and Recovery Strategies, Transportation Planning

16. eCommerce: Web Metrics, Web-page Visitor Analysis, Customizable Site Designs, Customer Recommendations, Online Advertisements, Online Trend Analysis, BLOG Analysis

Bibliography

Jeph Abara. Applying integer linear programming to the fleet assignment problem. *INTER-FACES*, 19(4), 1989. DOI: 10.1287/inte.19.4.20. 39, 41

Anurag Acharya, Mustafa Uysal, and Joel H. Saltz. Active disks: Programming model, algorithms, and evaluation. In *Proc. of Conference on Architectural Support for Programming Languages and Operating Systems*, March 1998. DOI: 10.1145/384265.291026. 73

G. Adomavicius and A. Tuzhilin. Toward the next generation of recommender systems: A survey of the state-of-the-art and possible extensions. *IEEE Transactions on Knowledge and Data Engineering*, 17(6):734–749, 2005. DOI: 10.1109/TKDE.2005.99. 23

Charu C. Aggarwal, Cecilia Procopiuc, Joel L. Wolf, Philip S. Yu, and Jong Soo Park. Fast algorithms for projected clustering. In *Proc. of the 1999 ACM SIGMOD Intl. Conf. on Management of Data*, pages 61–72, 1999. DOI: 10.1145/304182.304188. 16

Rakesh Agrawal and Ramakrishnan Srikant. Fast algorithms for mining association rules. In *Proc. of 1994 Int. Conf. Very large Data Bases*, pages 487–499, 1994. 18, 19, 20

Rakesh Agrawal, Tomasz Imielinski, and Arun Swami. Mining association rules between sets of items in large database. In *Proc. of the 1993 ACM SIGMOD Conference*, pages 207–216, 1993. DOI: 10.1145/170035.170072. 18, 19, 20

Rakesh Agrawal, Johannes Gehrke, Dimitrios Gunopulos, and Prabhakar Raghavan. Automatic subspace clustering of high dimensional data for data mining applications. In *Procs. of the 1998 ACM SIGMOD Intl. Conf. on Management of Data*, pages 94–105, 1998. DOI: 10.1145/276304.276314. 16

Dan Alcantara. *Efficient Hash Tables on the GPU*. Ph.D. thesis, University of California, Davis, 2011. 71

Monther Aldwairi, Thomas Conte, and Paul Franzon. Configurable string matching hardware for speeding up intrusion detection. *ACM SIGARCH Computer Architecture News*, 33(1), March 2005. DOI: 10.1145/1055626.1055640. 58

N. Alon. Spectral techniques in graph algorithms. *Lecture Notes in Computer Science 1380*, pages 206–215, 1998. DOI: 10.1007/BFb0054322. 46

David R. Anderson, Dennis J. Sweeney, Thomas A. Williams, Jeffrey D. Camm, and R. Kipp Martin. *Quantitative Methods for Business*, 11th ed. Cengage Learning, 2009. 9, 41

Alexandr Andoni and Piotr Indyk. Near-optimal hashing algorithms for approximate nearest neighbor in high dimensions. *Communications of the ACM*, 51(1):117–122, 2008. DOI: 10.1145/1327452.1327494. 18

K. R. Apt. *Principles of Constraint Programming*. Cambridge University Press, 2003. 40, 41

Chidanand Apte, Bing Liu, Edwin Pednault, and Padhraic Smyth. Business applications of data mining. *Communications of the ACM*, 45(8), August 2002. DOI: 10.1145/545151.545178. 9

Andrew P. Armacost, Cynthia Barnhart, Keith A. Ware, and Alysia M. Wilson. UPS Optimizes Its Air Network. *Interfaces*, 34(1), January-February 2004. DOI: 10.1287/inte.1030.0060. 2, 9, 41

Oliver Arnold, Sebastian Haas, Gerhard Fettweis, Benjamin Schlegel, Thomas Kissinger, Tomas Karnagel, and Wolfgang Lehner. Hashi: An application specific instruction set extension for hashing. In *Proc. of ADMS'14, Fifth International Workshop on Accelerating Data Management Systems Using Modern Processor and Storage Architectures*, August 2014. 66

Sunil Arya, David M. Mount, Nathan S. Netanyahu, Ruth Silverman, and Angela Y. Wu. An optimal algorithm for approximate nearest neighbor searching in fixed dimensions. *Journal of the ACM*, 45(6):891–923, 1998. DOI: 10.1145/293347.293348. 17, 18

Kubilay Atasu, Raphael Polig, Christoph Hagleitner, and Frederick R. Reiss. Hardware-accelerated regular expression matching for high-throughput text analytics. In *Proc. 23rd International Conference on Field Programmable Logic and Applications*, September 2013. DOI: 10.1109/FPL.2013.6645534. 58

Albert-Laszlo Barabasi. *Linked: How Everything Is Connected to Everything Else and What it Means for Business, Science, and Everyday Life*. Penguin Group, 2003. 44, 45, 46

Albert-Laszlo Barabasi and Eric Bonabeau. Scale-free networks. *Scientific American*, 288:60–69, May 2003. DOI: 10.1038/scientificamerican0503-60. 45, 46

Cynthia Barnhart, Ellis L. Johnson, George Nemhauser, Martin W. P. Savelbergh, and Pamela H. Vance. Branch-and-price: Column generation for solving huge integer programs. *Optimization*, 46(3):316–329, May-June 1998. DOI: 10.1287/opre.46.3.316. 41

Kim Baskerville and Maya Paczuski. Subgraph ensembles and motif discovery using an alternative heuristic for graph isomorphism. *Physical Review E 74*, pages 051903:1–11, 2006. 46

Frédéric Bastien, Pascal Lamblin, Razvan Pascanu, James Bergstra, Ian J. Goodfellow, Arnaud Bergeron, Nicolas Bouchard, David Warde-Farley, and Yoshua Bengio. Theano: new features and speed improvements. *CoRR*, abs/1211.5590, 2012. DOI: 10.1103/PhysRevE.74.051903. 61

C. Basu, H. Hirsh, and W. Cohen. Recommendation as classification: Using social and content-based information in recommendation. In *Proc. of the Fifteenth National Conference on Artificial Intelligence*, pages 714–720, July 1998. 23

Ron Bekkerman, Misha Bilenko, and John Langford, editors. *Scaling Up Machine Learning: Parallel and Distributed Approaches*. Cambridge University Press, 2011. 9, 53

Robert M. Bell, Yehuda Koren, and Chris Volinsky. All Together Now: A Perspective on the Netflix Prize. *Chance*, 23(1):24–29, 2010. DOI: 10.1080/09332480.2010.10739787. 9, 16, 18, 23

Y. Bengio. Learning deep architectures for AI. *Foundations and Trends in Machine Learning*, 2 (1), 2009. DOI: 10.1561/2200000006. 27, 59, 61

Yoshua Bengio, Ian J. Goodfellow, and Aaron Courville. Deep learning. Book in preparation for MIT Press, 2014. http://www.iro.umontreal.ca/~bengioy/dlbook. 13, 27, 34

Kristin Bennett and Colin Campbell. Support vector machines: Hype or hallelujah? *SIGKDD Explorations*, 2(2), 2000. DOI: 10.1145/380995.380999. 23, 25

Jon Louis Bentley. Multidimensional binary search trees used for associative searching. *Communications of the ACM*, 18(9):509–517, 1975. DOI: 10.1145/361002.361007. 17, 18

Jon Louis Bentley. Multidimensional divide and conquer. *Communications of the ACM*, 23(4): 214–229, 1980. DOI: 10.1145/358841.358850. 17, 18

James Bergstra, Olivier Breuleux, Frédéric Bastien, Pascal Lamblin, Razvan Pascanu, Guillaume Desjardins, Joseph Turian, David Warde-Farley, and Yoshua Bengio. Theano: a CPU and GPU math expression compiler. In *Proc. of the Python for Scientific Computing Conference (SciPy)*, June 2010. Oral Presentation. 61

M. W. Berry, S. T. Dumais, and G. W. O'Brien. Using linear algebra for intelligent information retrieval. Technical Report CS-94-270, Computer Science Department, University of Tennessee, Knoxville, December 1994. DOI: 10.1137/1037127. 48, 50

Michael W. Berry, Susan T. Dumais, and Gavin W. O'Brien. Using linear algebra for intelligent information retrieval. *SIAM Rev.*, 37(4), December 1995. DOI: 10.1137/1037127. 58

Kevin Beyer, Jonathan Goldstein, Raghu Ramakrishnan, and Uri Shaft. When is "nearest neighbor" meaningful? *Lecture Notes in Computer Science*, 1540:217–235, 1999. DOI: 10.1007/3-540-49257-7_15. 16, 18

Alina Beygelzimer, Sham Kakade, and John Langford. Cover trees for nearest neighbor. In *Procs. of the Twenty-Third International Conference (ICML 2006)*, pages 97–104, 2006. DOI: 10.1145/1143844.1143857. 18

Bharat Bhasker and K. Srikumar. *Recommender Systems in E-Commerce*. Cambridge University Press, 2010. 23

Indrajit Bhattacharya, Shantanu Godbole, Ajay Gupta, Ashish Verma, Jeff Achtermann, and Kevin English. Enabling analysts in managed services for CRM analytics. In *Proc. of the 2009 ACM KDD Intl. Conf. on Knowledge and Data Discovery*, 2009. DOI: 10.1145/1557019.1557136. 9

Jeff A. Bilmes. A gentle tutorial of the em algorithm and its application to parameter estimation for gaussian mixture and hidden markov models. Technical Report TR-97-021, International Computer Science Institute, April 1998. 15

Guy E. Blelloch. *Vector models for data-parallel computing*. MIT Press, Cambridge, MA, USA, August 1990. 67, 68

O. Boiman, E. Shechtman, and M. Irani. In defense of nearest-neighbor based image classification. In *Proc. of the IEEE Conference on Computer Vision and Pattern Recognition (CVPR 2008)*, pages 1–8, June 2008. DOI: 10.1109/CVPR.2008.4587598. 16

Rajesh Bordawekar and Daniel Beece. Financial risk modeling on low-power accelerators: xperimental performance evaluation of tk1 with fpga, March 2015. Nvidia Global Technology Conference. 63, 64

Rajesh Bordawekar, Bob Blainey, Chidanand Apte, and Michael McRoberts. Analyzing analytics, part 1: A survey of business analytics models and algorithms. Technical Report RC25186, IBM T. J. Watson Research Center, July 2011. DOI: 10.1145/2590989.2590993. 8

Jacir L. Bordim, Yasuaki Ito, and Koji Nakano. Accelerating the cky parsing using fpgas. *IEICE Trans. Inf. and Syst.*, E86-D(5), May 2003. DOI: 10.1007/3-540-36265-7_5. 57, 58

Bernhard E. Boser, Isabelle M. Guyon, and Vladimir Vapnik. A training algorithm for optimal margin classifiers. In *Fifth Annual Workshop on Computational Learning Theory*, pages 144–152, 1992. DOI: 10.1145/130385.130401. 23, 25

George Box and Gwilym Jenkins. *Time Series Analysis: Forecasting and Control*. Holden-Day, 1976. 31, 32

S. Boyd, N. Parikh, E. Chu, B. Peleato, and J. Eckstein. Distributed optimization and statistical learning via the alternating direction method of multipliers. *Foundations and Trends in Machine Learning*, 3(1):1–122, 2011. DOI: 10.1561/2200000016. 53

Stephen Boyd and Lieven Vandenberghe. *Convex Optimization*. Cambridge University Press, 2009. 41

Phelim P. Boyle. Options: A monte carlo approach. *Journal of Financial Economics*, 4:323–338, 1977. DOI: 10.1016/0304-405X(77)90005-8. 36, 37

J.S. Breese, D. Heckerman, and C. Kadie. Empirical analysis of predictive algorithms for collaborative filtering. In *Proc. of the Fourteenth Conference on Uncertainty in Artificial Intelligence*, July 1998. 23

L. Breiman, J. H. Friedman, R. A. Olshen, and C. J. Stone. *Classification and Regression Trees*. Wadsworth, Belmont, CA, 1984. 29, 30

Coen Bron and Joep Kerbosch. Algorithm 457: Finding all cliques of an undirected graph. *Communications of the ACM*, 16(9):575–577, 1973. DOI: 10.1145/362342.362367. 46

Kurt Bryan and Tanya Leise. The $25,000,000,000 eigenvector: The linear algebra behind google. *SIAM Rev.*, 48(3), March 2006. DOI: 10.1137/050623280. 67, 68

Christopher Burges. A tutorial on support vector machines for pattern recognition. *Data Mining and Knowledge Discovery*, 2:121–167, 1998. DOI: 10.1023/A:1009715923555. 23, 25

Business Application Research Center. The BI Verdict. www.bi-verdict.com. 43, 44

Business Week. Math Will Rock Your World, January 2006. 9

John Canny, David Hall, and Dan Klein. A multi-teraflop constituency parser using gpus. In *Proc. 2013 Conference on Empirical Methods in Natural Language Processing*, October 2013. 57, 58

Richard Cantor, Frank Packer, and Kevin Cole. Split ratings and the pricing of credit risk. Technical Report Research Paper No. 9711, Federal Reserve Bank of New York, March 1997. DOI: 10.3905/jfi.1997.408217. 9

G. A. Cecchi, A. Ma'ayan, A. R. Rao, J. Wagner, R. Iyengar, and G. Stolovitzky. Ordered cyclic motifs contributes to dynamic stability in biological and engineered networks. *Proc. of the National Academy of Sciences*, 105:19235–19240, 2008. DOI: 10.1073/pnas.0805344105. 46

Deepayan Chakrabarti and Christos Faloutsos. Graph mining: Laws, generators, and algorithms. *ACM Computing Surveys*, 38(2), March 2006. DOI: 10.1145/1132952.1132954. 44, 45, 46

Don Chamberlin. *A Complete Guide to DB2 Universal Database*. Morgan Kaufmann Publishers, 1998. 44

A. Chaturvedi, P. E. Green, and J. D. Caroll. K-modes clustering. *Journal of Classification*, 18 (1):35–55, 2001. DOI: 10.1007/s00357-001-0004-3. 15, 16

S. Chaudhuri and U. Dayal. An overview of data warehousing and OLAP technology. *SIGMOD Record*, 26:65–74, 1997. DOI: 10.1145/248603.248616. 41, 42, 43, 44

Tianshi Chen, Yunji Chen, Marc Duranton, Qi Guo, Atif Hashmi, Mikko Lipasti, Andrew Nerei, Shi Qiu, Michele Sebag, and Olivier Temam. Benchnn: On the broad potential application scope of hardware neural network accelerators. In *Proc. International Symposium on Workload Characterization (IISWC'12)*, 2012. DOI: 10.1109/IISWC.2012.6402898. 62, 76

Tianshi Chen, Zidong Du, Ninghui Sun, Jia Wang, Chengyong Wu, Yunji Chen, and Olivier Temam. Diannao: A small-footprint high-throughput accelerator for ubiquitous machine-learning. In *Proc. of the 19th ACM International Conference on Architectural Support for Programming Languages and Operating Systems (ASPLOS'14)*, 2014a. DOI: 10.1145/2541940.2541967. 62, 76

Yunji Chen, Tao Luo, Shaoli Liu, Shijin Zhang, Liqiang He, Jia Wang, Ling Li, Tianshi Chen, Zhiwei Xu, Ninghui Sun, and Olivier Temam. Dadiannao: A machine-learning supercomputer. In *Proc. of the 47th IEEE/ACM International Symposium on Microarchitecture (MICRO'14)*, 2014b. DOI: 10.1109/MICRO.2014.58. 61, 62, 76

Sharan Chetlur, Cliff Woolley, Philippe Vandermersch, Jonathan Cohen, John Tran, Bryan Catanzaro, and Evan Shelhamer. cudnn: Efficient primitives for deep learning. *CoRR*, abs/1410.0759, 2014. 60, 61

Trishul Chilimbi, Yutaka Suzue, Johnson Apacible, and Karthik Kalyanaraman. Project adam: Building an efficient and scalable deep learning training system. In *Proc. of the 11th USENIX Symposium on Operating Systems Design and Implementation*, October 2014. 60, 61, 75, 76

Vinay K. Chippa, Srimat T. Chakradhar, Kaushik Roy, and Anand Raghunathan. Analysis and characterization of inherent application resilience for approximate computing. In *Proc. of the 50th Annual Design Automation Conference, DAC'13*, 2013. DOI: 10.1145/2463209.2488873. 73, 76

Tom Chiu, DongPing Fang, John Chen, Yao Wang, and Christopher Jeris. A robust and scalable clustering algorithm for mixed type attributes in large database environment. In *Procs. of the 2001 ACM KDD Intl. Conf. on Knowledge and Data Discovery*, pages 263–268, 2001. DOI: 10.1145/502512.502549. 16

Henning Marxen Christian de Schryver and Daniel Schmidt. Hardware accelerators for financial mathematics - methodology, results and benchmarks. In *Proc. Young Researcher's Symposium (YRS)*, February 2011. 62, 64

I-Hsin Chung, Tara N. Sainath, Bhuvana Ramabhadran, Michael Picheny, John Gunnels, Vernon Austel, Upendra Chauhari, and Brian Kingsbury. Parallel deep neural network training for big data on blue gene/q. In *Proc. of SC14: International Conference for High Performance Computing, Networking, Storage and Analysis*, November 2014. DOI: 10.1109/SC.2014.66. 60

P. Ciaccia, M. Patella, and P. Zezula. M-tree: An efficient access method for similarity search in metric spaces. In *Proc. of the 23rd VLDB Conference*, 1997. 18

M. Claypool, A. Gokhale, and T. Miranda. Combining content-based and collaborative filters in an online newspaper. In *Proc. of the SIGIR-99 Workshop on Recommender Systems: Algorithms and Evaluations*, 1999. DOI: 10.1016/j.ijar.2010.04.001. 23

E. F. Codd, S. B. Codd, and C. T. Salley. Providing OLAP (On-line Analytics Processing) to user-analysts: An IT mandate, 1993a. DOI: 10.1016/j.sbspro.2014.07.171. 41, 43

E. F. Codd, S. B. Codd, and C. T. Salley. Beyond decision support. *Computer World*, 27, July 1993b. DOI: 10.1016/0360-8352(88)90043-5. 41, 43

COIN-OR Foundation. COmputational INfrastructure for Operations Research, 2011. http://www.coin-or.org. 41

Ronan Collobert, Koray Kavukcuoglu, and Clément Farabet. Torch7: A matlab-like environment for machine learning. In *BigLearn, NIPS Workshop*, 2011. 61

Jack G. Conrad, Khalid Al-Kofahi, Ying Zhao, and George Karypis. Effective document clustering for large heterogeneous law firm collections. In *10th International Conference on Artificial Intelligence and Law (ICAIL)*, pages 177–187, 2005. DOI: 10.1145/1165485.1165513. 34, 35

World Wide Web Consortium. XML Path Language (XPath) 2.0, W3C Recommendation, January 23, 2007. www.w3.org, a. 46

World Wide Web Consortium. XQuery 1.0: An XML Query Language, W3C Recommendation, January 23, 2007. www.w3.org, b. 46

Cork Constraint Computation Centre, University College Cork. CSP tutorial, 2011. http://4c.ucc.ie/web/outreach/tutorial.html. 40, 41

Corinna Cortes and Vladimir Vapnik. Support vector networks. *Machine Learning*, 20:273–297, 1995. DOI: 10.1007/BF00994018. 23, 25

Tom Van Court and Martin C. Herbordt. Families of fpga-based accelerators for approximate string matching. *Microprocess Microsyst.*, 31(2), March 2007. DOI: 10.1016/j.micpro.2006.04.001. 58

R. Couture and P. L'Ecuyer. On the lattice structure of certain linear congruential sequences related to awc/swb generators. *Mathematics of Computation*, 62:799–808, 1994. DOI: 10.1090/S0025-5718-1994-1220826-X. 37

Carroll Croarkin and Paul Tobias. NIST/SEMATECH e-Handbook of Statistical Methods, 2011. 31, 32, 33

Brendan Cronin. *Hardware Acceleration of Network Intrusion Detection and Prevention.* Ph.D. thesis, Dublin City University, January 2014. 58

Peter Crosbie and Jeff Bohn. Modeling default risk, December 2003. Moody's KMV. 2, 9

Zefu Dai, Nick Ni, and Jianwen Zhu. A 1 cycle-per-byte xml parsing accelerator. In *Proc. FPGA'10*, February 2010. DOI: 10.1145/1723112.1723148. 58

G. B. Dantzig. *Linear Programming and Extensions.* Princeton University Press, Princeton, NJ, 1963. DOI: 10.1145/997817.997857. 39, 41

Koustuv Dasgupta, Rahul Singh, Balaji Viswanathan, Dipanjan Chakraborty, Sougata Mukherjea, Amit Anil Nanavati, and Anupam Joshi. Social ties and their relevance to churn in mobile telecom networks. In *EDBT 2008, 11th International Conference on Extending Database Technology*, pages 668–677, March 2008. DOI: 10.1145/1353343.1353424. 9, 45, 46

Mayur Datar, Nicole Immorlica, Piotr Indyk, and Vahab S. Mirrokni. Locality-sensitive hashing scheme based on p-stable distributions. In *Symposium on Computational Geometry*, pages 253–262, 2004. DOI: 10.1145/997817.997857. 18

T. Davenport and J. Harris. *Competing on Analytics, The New Science of Winning.* Harvard Business School Press, 2007. 2, 5, 9, 20

T. Davenport, J. Harris, and R. Morison. *Analytics at Work, Smarter Decisions, Better Results.* Harvard Business School Press, 2010. 5, 9, 20

Angelique Davi, Dominique Haughton, Nada Nasr, Gaurav Shah, Maria Skaletsky, and Ruth Spack. A review of two text-mining packages: Sas textmining and wordstat. *The American Statistician*, 59(1):89–103, February 2005. DOI: 10.1198/000313005X22987. 33, 35

Christian de Schryver, Ivan Shcherbakov, Frank Kienle, Norbert Wehn, Henning Marxen, Anton Kostiuk, and Ralf Korn. An energy efficient fpga accelerator for monte carlo option pricing with the heston model. In *Proc. 2011 International Conference on Reconfigurable Computing and FPGAs*, 2011. DOI: 10.1109/ReConFig.2011.11. 64

Jeff Dean. Large scale deep learning, March 2015. Nvidia Global Technology Conference Keynote. 60, 61, 76

Jeffrey Dean and Sanjay Ghemawat. Mapreduce: a flexible data processing tool. *Communications of the ACM*, 53(1), 2010. DOI: 10.1145/1629175.1629198. 8

Rina Dechter. *Constraint Processing.* Morgan Kaufmann, 2003. 40, 41

Scott Deerwester, Susan T. Dumais, George W. Furnas, Thomas K. Landauer, and Richard Harshman. Indexing by latent semantic analysis. *Journal of the American Society For Information Science*, 41:391–407, 1990. DOI: 10.1002/(SICI)1097-4571(199009)41:6%3C391::AID-ASI1%3E3.0.CO;2-9. 34, 35

Christos Delivirias. Case studies in acceleration of hestons stochastic volatility financial engineering model: Gpu, cloud and fpga implementations, 2012. M.Sc. in Operational Research Thesis. 62, 64

A. P. Dempster, N. M. Laird, and D. B. Rubin. Maximum likelihood from incomplete data via the em algorithm. *Journal of the Royal Statistical Society. Series B (Methodological)*, 39(1):1–38, 1977. 15, 16

Sarang Dharmapurikar, Praveen Krishnamurthy, Todd S. Sproull, and John W. Lockwood. Deep packet inspection using parallel Bloom filters. *IEEE Micro*, 24(1):52–61, January 2004. DOI: 10.1109/MM.2004.1268997. 71, 75

Inderjit Dhillon, Yuqiang Guan, and Brian Kulis. A unified view of kernel k-means, spectral clustering, and graph cuts. Technical Report TR-04-25, Department of Computer Science, University of Texas at Austin, February 2005. 46

Chris Ding, Tao Li, Wei Peng, and Haesum Park. Orthogonal nonnegative matrix tri-factorizations for clustering. In *Proc. of KDD'06*, pages 126–134, 2006. DOI: 10.1145/1150402.1150420. 35

Chris Ding, Tao Li, and Michael I. Jordan. Nonnengative matrix factorization for combinatorial optimization: Spectral clustering, graph matching, and clique finding. In *2008 Eighth IEEE International Conference on Data Mining*, pages 183–192, 2008. DOI: 10.1109/ICDM.2008.130. 46

Paul Dlugosch, Dave Brown, Paul Glendenning, Michael Leventhal, and Harold Noyes. An efficient and scalable semiconductor architecture for parallel automata processing. *IEEE Transactions on Parallel and Distributed Systems*, 25(12), December 2014. DOI: 10.1109/TPDS.2014.8. 58

Goutam Dutta and Robert Fourer. A survey of mathematical programming applications in integrated steel plants. *Manufacturing and Service Operations Management*, 3(4):387–400, Fall 2001. DOI: 10.1287/msom.3.4.387.9972. 38, 41

Wayne W. Eckerson. Beyond Reporting: Delivering Insights with Next-Generation Analytics, 2009. TDWI Best Practices Report. 9

Jeffrey L. Elman. Finding structure in time. *Cognitive Science*, 14(2):179–211, 1990. DOI: 10.1207/s15516709cog1402_1. 28

Michael Falk, Frank Marohn, Rene Michel, Daniel Hofmann, and Maria Macke. A first course on time series analysis—examples with SAS, 2006. 30, 31, 32

C. Farabet, B. Martini, P. Akselrod, S. Tlay, Y. LeCun, and E. Culurciello. Hardware accelerated convolutional neural networks for synthetic vision systems. In *Proc. International Symposium on Circuits and Systems (ISCAS'10)*, 2010. DOI: 10.1109/ISCAS.2010.5537908. 61, 62, 76

Amer Farroukh, Mohammad Sadoghi, and Hans-Arno Jacobsen. Towards vulnerability-based intrusion detection with event processing. In *Proc. of the 5th ACM international conference on Distributed event-based system*, DEBS'11, pages 171–182, New York, New York, USA, 2011. ACM. DOI: 10.1145/2002259.2002284. 75

Ingo Feinerer, Kurt Hornik, and David Meyer. Text Mining Infrastructure in R. *Journal of Statistical Software*, 25(5), March 2008. DOI: 10.18637/jss.v025.i05. 33, 35

Daniel J. Feldman. Netezza Performance Architecture, June 2013. Keynote Presentation at the DaMoN'13 Workshop. 64, 66, 75

David Ferrucci, Eric Brown, Jennifer Chu-Carroll, James Fan, David Gondek, Aditya A. Kalyan-pur, Adam Lally, J. William Murdock, Eri Nyberg, John Prager, Nico Schlaefer, and Chris Welty. Building Watson: An Overview of the DeepQA Project. *AI Magazine*, 59(Fall), 2010. 9, 47

George S. Fishman. *Monte Carlo Concepts, Algorithms, and Applications*. Springer Verlag, 1996. DOI: 10.1007/978-1-4757-2553-7. 36, 37

Blake Fitch. Active storage: Exploring a scalable, compute-in-storage model by extending the blue gene/q architecture with integrated non-volatile memory. In *Proc. of ADMS'13, Fourth International Workshop on Accelerating Data Management Systems Using Modern Processor and Storage Architectures*, August 2013. 70, 73, 75

Tristan Fletcher. Support vector machines explained, March 2009. 24

Tim Foley and Jeremy Sugerman. Kd-tree acceleration structures for a gpu raytracer. *Graphics Hardware*, 2005. DOI: 10.1145/1071866.1071869. 71, 75

Peter Forsyth. An introduction to computational finance without agonizing pain, December 2014. 62, 64

Jeremy Fowers, Kalin Ovtcharov, Karin Strauss, Eric Chung, and Greg Stitt. A high memory bandwidth fpga accelerator for sparse matrix-vector multiplication. In *Proc. of International Symposium on Field-Programmable Custom Computing Machines*, May 2014. DOI: 10.1109/FCCM.2014.23. 72, 76

Regis Fricker. A true story: Gpu in production for intraday risk calculations, March 2015. *Nvidia Global Technology Conference.* 64

J. H. Friedman, J. L. Bentley, and R. A. Finkel. An algorithm for finding best matches in logarithmic expected time. *ACM Transactions on Mathematical Software*, 3(3):209–226, September 1977. DOI: 10.1145/355744.355745. 17, 18

Zhisong Fu, Bryan Thompson, and Michael Personick. Mapgraph: A high level api for fast development of high performance graph analytics on gpus. In *Proc. of GRADES 2014*, June 2014. DOI: 10.1145/2621934.2621936. 67, 68

Kunihiko Fukushima. Neocognitron: A self-organizing neural network model for a mechanism of pattern recognition unaffected by shift in position. *Biological Cybernetics*, 36(4):193–202, 2013. DOI: 10.1007/BF00344251. 27

Thomas Gartner. A survey of kernels for structured data. *SIGKDD Explorations*, 2003. DOI: 10.1145/959242.959248. 35

Bugra Gedik, Henrique Andrade, Kun-Lung Wu, Philip S. Yu, and Myungcheol Doo. Spade: The system s declarative stream processing engine. In *Proc. of the 2008 ACM SIGMOD International Conference on Management of Data*, SIGMOD '08, 2008. DOI: 10.1145/1376616.1376729. 65, 66

Mike Giles. Computational finance on gpus, 2010. DOI: 10.1145/2535557.2535564. 63, 64

Philip E. Gill, Walter Murray, Michael A. Saunders, J. A. Tomlin, and Margaret H. Wright. On projected newton barrier methods for linear programming and an equivalence to karmarkar's projective method. *Mathematical Programming: Series A and B*, 36(2), November 1986. DOI: 10.1007/BF02592025. 41

William T. Glaser, Timothy B. Westergren, Jeffrey P. Stearns, and Jonathan M. Kraft. Consumer item matching method and system, 2002. United States Patent: US 7,003,515 B1. 9

Paul Glasserman. *Monte Carlo Methods in Financial Engineering.* Springer, 2003. DOI: 10.1007/978-0-387-21617-1. 36, 37

D. Goldberg, D. Nichols, B. Oki, and D. Terry. Using collaborative filtering to weave an information tapestry. *Communications of the ACM*, 35(12):61–70, 1992. DOI: 10.1145/138859.138867. 23

N. Good, J. B. Schafer, J. A. Konstan, A. Borchers, B. Sarwar, and J. Herlocker. Combining collaborative filtering with personal agents for better recommendations. In *Proc. of the Sixteenth National Conference on Artifical Intelligence*, pages 439–446, July 1999. 23

Erica Goode. Sending the police before there's a crime. *The New York Times*, August 16, 2011. 2, 9

Nick Gould and Philippe Toint. A quadratic programming page. `http://www.numerical.rl.ac.uk/qp/qp.html`. 41

Scott Grauer-Gray, William Killian, Robert Searles, and John Cavazos. Accelerating financial applications on the gpu. In *Proc. of GPGPU-6*, March 2013. DOI: 10.1145/2458523.2458536. 64

Jim Gray, Surajit Chaudhuri, Adam Bosworth, Andrew Layman, Don Reichart, Murali Venka-trao, Frank Pellow, and Hamid Pirahesh. Data cube: A relational aggregation operator gener-alizing group-by, cross-tab, and sub-totals. *Data Mining and Knowledge Discovery*, 1(1):29–53, 1997. DOI: 10.1023/A:1009726021843. 42, 43, 44

H. J. Greenberg. *Myths and Counter-examples in Mathematical Programming*. INFORMS Com-puting Society, `http://glossary.computing.society.informs.org`, February 2010. (ongoing, first posted October 2008). 38, 41

Joshua A. Grochow and Manolis Kellis. Network motif discovery using subgraph enumeration and symmetry-breaking. In *RECOMB, 2007*, 2007. DOI: 10.1007/978-3-540-71681-5_7. 46

Suyog Gupta and Kailash Gopalakrishnan. Revisiting stochastic computation: Approximate es-timation of machine learning kernels. In *Proc. of WACAS 2014*, 2014. 73, 76

Suyog Gupta, Ankur Agrawal, Kailash Gopalakrishnan, and Pritish Narayanan. Deep learning with limited numerical precision, February 2015. arXic:1502.02551v1. 50, 73, 76

Gurobi Optimization Inc. Gurobi Optimizer 4.5. `www.gurobi.com`. 41

Isabelle Guyon. SVM application list, 2006. `http://www.clopinet.com/isabelle/Projects/SVM/applist.html`. 23, 25

Mark Hall, Eibe Frank, Geoffrey Holmes, Bernhard Pfahringer, Peter Reutemann, and Ian H. Witten. The WEKA Data Mining Software: An Update. *SIGKDD Explorations*, 11(1), 2009. `www.cs.waikato.ac.nz/ml/weka`. DOI: 10.1145/1656274.1656278. 9, 14, 16, 20, 25, 28, 30, 33

J. H. Halton. A retrospective and prospective survey of the monte carlo method. *SIAM Review*, 12(1):1–63, 1970. DOI: 10.1137/1012001. 36, 37

J. Han, J. Pei, and Y. Yin. Mining frequent patterns without candidate generation. In *Proc. of the 2000 ACM SIGMOD Intl. Conf. on Management of Data*, 2000. DOI: 10.1145/342009.335372. 20

Jiawei Han and Micheline Kamber. *Data Mining: Concepts and Techniques*. Morgan Kaufmann Publishers, 2006. 7, 9, 14, 15, 16, 18, 20, 29, 30

Jiawei Han, Jain Pei, Yiwen Yin, and Runying Mao. Mining frequent patterns without candidate generation: A frequent-pattern tree approach. *Data Mining and Knowledge Discovery*, 8:53–87, 2004. DOI: 10.1023/B:DAMI.0000005258.31418.83. 20

Awni Y. Hannun, Carl Case, Jared Casper, Bryan C. Catanzaro, Greg Diamos, Erich Elsen, Ryan Prenger, Sanjeev Satheesh, Shubho Sengupta, Adam Coates, and Andrew Y. Ng. Deep speech: Scaling up end-to-end speech recognition. *CoRR*, abs/1412.5567, 2014. 61

Frank Harary. *Graph Theory*. Addison-Wesley, Reading, MA, 1969. 46

J. A. Harding, M. Shahbaz, Srinivas, and A. Kusiak. Data mining in manufacturing: A review. *Journal of Manufacturing Science and Engineering*, 128(4), 2006. DOI: 10.1115/1.2194554. 9

Venky Harinarayan, Anand Rajaraman, and Jeffrey D. Ullman. Implementing data cubes efficiently. In *Procs. of the 1996 ACM SIGMOD International Conference on Management of Data (SIGMOD96)*, pages 205–216, 1996. DOI: 10.1145/233269.233333. 42, 44

Pawan Harish and P. J. Narayanan. Accelerating large graph algorithms on the gpu using cuda. In *Proc. 14th International Conference on High Performance Computing, HiPC'07*, December 2007. DOI: 10.1007/978-3-540-77220-0_21. 67, 68

Pawan Harish, Vibhav Vineet, and P. J. Narayanan. Large graph algorithms for massively multithreaded architectures. Technical Report IIIT/TR/2009/74, Indian Institute of Information Technology, 2009. 68

J. A. Hartigan and M. A. Wong. Algorithm AS 136: A K-Means Clustering Algorithm. *Journal of the Royal Statistical Society, Series C (Applied Statistics)*, 28(1):100–108, 1979. DOI: 10.2307/2346830. 14, 16

Johann Hauswald, Michael A. Laurenzano, Yunqi Zhang, Cheng Li, Austin Rovinski, Arjun Khurana, Ronald G. Dreslinski, Trevor Mudge, Vinicius Petrucci1, Lingjia Tang, and Jason Mars. Sirius: An open end-to-end voice and vision personal assistant and its implications for future warehouse scale computers. In *Proc. of Conference on Architectural Support for Programming Languages and Operating Systems*, March 2015. DOI: 10.1145/2775054.2694347. 69

Bingsheng He, Naga K. Govindaraju, Qiong Luo, and Burton Smith. Efficient gather and scatter operations on graphics processors. In *Proc. of the 2007 ACM/IEEE Conference on Supercomputing*, pages 46:1–46:12, November 2007. DOI: 10.1145/1362622.1362684. 67, 68

Donald O. Hebb. *The Organization of Behavior: A Neuropsychological Theory*. Wiley, New York, 1949. 26, 28

Robert Henschel, Abhik Seal, Jeremy J. Yang, David J. Wild, Ying Ding, Abhinav Thota, Scott Michael, Matt Gianni, and Jim Maltby. Applications of the yarcdata urika in drug discovery and healthcare. In *Proc. 2014 Cray Users Group Conference*, May 2014. 67, 68

Rick Hetherington. SPARC at Oracle: Vectoring Processor Architecture at the Database, September 2015. Keynote presentation at the ADMS'15 Workshop. 66

G. Hinton, L. Deng, D. Yui, G. Dahli, A. Mohamedi, N. Jaitly, A. Seniori, V. Vanhoucke, P. Nguyeni, T. N. Sainath, , and B. Kingsbury. Deep neural networks for acoustic modeling in speech recognition. *IEEE Signal Processing Magazine*, 29(6):82–97, 2012. DOI: 10.1109/MSP.2012.2205597. 59, 61

Jochen Hipp, Ulrich Guntzer, and Gholamreza Nakhaeizadeh. Algorithms for association rule mining: A general survey and comparison. *SIGKDD Explorations*, 2(1):58–64, 2000. DOI: 10.1145/360402.360421. 19, 20

Ngoc-Diep Ho. *Non-negative Matrix Factorization Algorithms and Applications*. Ph.D. thesis, Universite catholique de Louvain, June 2008. 35

Karla Hoffman. Combinatorial optimization: History and future challenges. *Journal of Applied and Computational Mathematics*, 124:341–360, 2000. DOI: 10.1016/S0377-0427(00)00430-1. 40, 41

A. Holder, editor. *Mathematical Programming Glossary*. INFORMS Computing Society, `http://glossary.computing.society.informs.org`, 2006–2008. Originally authored by Harvey J. Greenberg, 1999-2006. 38, 40, 41

John J. Hopfield. Neural networks and physical systems with emergent collective computational ability. *Proc. of the National Academy of Sciences of the USA*, 79(8):2554–2558, April 1982. DOI: 10.1073/pnas.79.8.2554. 27, 28

Chih-Wei Hsu, Chih-Chung Chang, and Chih-Jen Lin. A proactical guide to support vector classification, April 2010. 23, 25

Y. Hu, Y. Koren, and C. Volinsky. Collaborative filtering for implicit feedback datasets. In *Proc. of the Eighth IEEE International Conference on Data Mining*, pages 263–272, December 2008. DOI: 10.1109/ICDM.2008.22. 22, 23

IBM Corp. IBM Cognos, a. `www.ibm.com/software/data/cognos`. 44

IBM Corp. IBM ILOG CPLEX Optimization Studio Documentation, b. 39, 41

IBM Corp. IBM InfoSphere Warehouse, c. `www.ibm.com/software/data/infosphere/warehouse`. 16, 20, 28

IBM Corp. Real-time Analysis for Intensive Care, 2011. InfoSphere Streams for Smarter Healthcare White Paper. 9

IBM Institute for Business Value. From social media to Social CRM, February 2011. IBM Global Business Services Executive Report. 9

IBM Netezza. Netezza Data Warehouse Appliances. www.netezza.com. 44

IBM SPSS. SPSS Modeler, 2010a. 16, 25, 28, 30, 33, 46

IBM SPSS. SPSS Statistics 19, 2010b. 14

IBM SPSS. IBM SPSS Text Analytics for Surveys, 2010c. 33, 35

P. Indyk and R. Motwani. Approximate nearest neighbors: Towards removing the curse of dimensionality. In *Proc. of 30th Sympo. on Theory of Computing*, pages 604–613, 1998. DOI: 10.1145/276698.276876. 18

Piotr Indyk. Nearest neighbors in high-dimensional spaces, 2004. DOI: 10.1201/9781420035315.ch39. 16

Hiroshi Inoue, Takao Moriyama, Hideaki Komatsu, and Toshio Nakatani. Aa-sort: A new parallel sorting algorithm for multi-core simd processors. In *Proc. of the 16th International Conference on Parallel Architecture and Compilation Techniques*, 2007. DOI: 10.1109/PACT.2007.4336211. 65, 72, 76

Intel Corp. Analyzing business as it happens, April 2011. 44

Arpith Jacob and Maya Gokhale. Language classification using n-grams accelerated by fpga-based bloom filters. In *Proc. HPRCTA'07*, November 2007. DOI: 10.1145/1328554.1328564. 76

Ali Jadbabaie, Jie Lin, and A. Stephen Morse. Coordination of groups of mobile autonomous agents using nearest neighbor rules. *IEEE Transactions on Automatic Control*, 40(6), June 2003. DOI: 10.1109/TAC.2003.817537. 16, 18

Jedox AG. Palo OLAP Server for Excel. www.jedox.com/en/products/palo-for-excel/palo-for-excel.html. 44

Yangqing Jia, Evan Shelhamer, Jeff Donahue, Sergey Karayev, Jonathan Long, Ross B. Girshick, Sergio Guadarrama, and Trevor Darrell. Caffe: Convolutional architecture for fast feature embedding. *CoRR*, abs/1408.5093, 2014. DOI: 10.1145/2647868.2654889. 61

Qiwei Jin, Wayne Luk, and David B. Thomas. On comparing financial option price solvers on fpga. In *Proc. IEEE 19th Annual International Symposium on Field-Programmable Custom Computing Machines, FCCM 2011*, May 2011. DOI: 10.1109/FCCM.2011.30. 64

T. Joachims. Text categorization with support vector machines: Learning with many relevant features. In *European Conference on Machine Learning (ECML)*, pages 137–142, Berlin, 1998. Springer. DOI: 10.1007/BFb0026683. 34, 35

Mark Johnson. Parsing in parallel on multiple cores and gpus. In *Proc. Australasian Language Technology Association Workshop*, December 2011. DOI: 10.1109/HPCC.2011.74. 57, 58

M. I. Jordan. Attractor dynamics and parallelism in a connectionist sequential machine. In *Proc. of the Eighth Annual Conference of the Cognitive Science Society*, pages 531–546, 1986. DOI: 10.1109/HPCC.2011.74. 28

John Joyce. Pandora and the Music Genome Project. *Scientific Computing*, 23(10):40–41, September 2006. 9, 16, 18

Alexander Kaganov, Asif Lakhany, and Paul Chow. Fpga acceleration of multifactor cdo pricing. *ACM Trans. Reconfigurable Technol. Syst.*, 4(2), May 2011. DOI: 10.1145/1968502.1968511. 63, 64

Khushboo Kanjani. Parallel non negative matrix factorization for document clustering, 2007. 35

N. Karmarkar. A new polynomial-time algorithm for linear programming. *Combinatorica*, 4: 373–395, 1984. DOI: 10.1007/BF02579150. 39

G. V. Kass. An exploratory technique for investigating large quantities of categorical data. *Applied Statistics*, 29(2):119–127, 1980. DOI: 10.2307/2986296. 30

S. Kirkpatrick, C. D. Gelatt, and M. P. Vecchi. Optimization by simulated annealing. *Science*, 220(4598):671–680, 1983. DOI: 10.1126/science.220.4598.671. 41

Graham Kirsch. Active memory: Micron's yukon. In *Proc. of the International Parallel and Distributed Processing Symposium*, April 2003. DOI: 10.1109/IPDPS.2003.1213195. 70, 73, 76

Jon Kleinberg and Andrew Tomkins. Applications of linear algebra in information retrieval and hypertext analysis. In *PODS'99, Proc. of the eighteenth ACM SIGMOD-SIGACT-SIGART Symposium on Principle of Database Systems*, pages 185–193, June 1999. DOI: 10.1145/303976.303995. 50

M. Klemettinen, H. Mannila, P. Ronkainen, H. Toivonen, and A. Verkamo. Finding interesting rules from large sets of discovered association rules. In *Proc. of the 3rd Intl. Conf. on Information and Knowledge Management*, 1994. DOI: 10.1145/191246.191314. 19, 20

Masayoshi Kobayashi, Tutomu Murase, and Atsushi Kuriyama. A longest prefix match search engine for multi-gigabit ip processing. In *Proc. of the International Conference on Communications*, pages 1360–1364. IEEE, 2000.

Teuvo Kohonen. *Self-Organization Maps*. Springer-Verlag, 1997. 28

Teuvo Kohonen and Timo Honkela. Kohonen network, 2007. *Scholarpedia*, 2(1):1568. 28

Andrew Kopser and Dennis Vollrath. Overview of the Next Generation Cray XMT. In *Proc. 2014 Cray Users Group Conference*, April 2012. 66, 68

Y. Koren, R. Bell, and C. Volinsky. Matrix factorization techniques for recommender systems. *IEEE Computer*, 42(8):30–37, 2009. DOI: 10.1109/MC.2009.263. 23

Ralf Korn. Computational problems in pricing, risk, and asset management in banks and insurance companies, September 2014. FPL 2014 Tutorial: Reconfigurable Architectures in Finance. 62, 64

S. B. Kotsiantis. Supervised machine learning: A review of classification techniques. *Informatica*, 31:249–268, 2007. 29, 30

Valdis Krebs. Circular CDOs: Contagion in the financial industry. orgnet.com/cdo.html. 46

Hans-Peter Kriegel, Peer Kroger, and Arthur Zimek. Clustering high-dimensional data: A survey on subspace clustering, pattern-based clustering, and correlation clustering. *ACM Trans. Knowl. Discov. Data*, 3:1:1–1:58, March 2009. DOI: 10.1145/1497577.1497578. 14

David Kriesel. A brief introduction to neural networks, 2007. http://www.dkriesel.com. 25, 26, 28

Alex Krizhevsky. One weird trick for parallelizing convolutional neural networks, April 2014. arXiv:1404.5997v2. 60

J. Kruskal. Multidimensional scaling by optimizing goodness of fit to a non metric hypothesis. *Psychometrika*, 29(1):1–27, 1964. DOI: 10.1007/BF02289565. 14

Volodymyr Kysenko, Karl Rupp, Oleksandr Marchenko, Siegfried Selberherr, and Anatoly Anisimov. Gpu-accelerated non-negative matrix factorization for text mining. In *Proc. of the 17th International Conference on Applications of Natural Language Processing and Information Systems*, 2012. DOI: 10.1007/978-3-642-31178-9_15. 58

Susana Ladra, Oscar Pedreira, Jose Duato, and Nieves R. Brisaboa. Exploiting simd instructions in current processors to improve classical string algorithms. In *Proc. of the 16th East European Conference on Advances in Databases and Information Systems*, ADBIS'12, 2012. DOI: 10.1007/978-3-642-33074-2_19. 58

Tirthankar Lahiri. Oracle's in-memory data management strategy: In-memory in all tiers, and for all workloads. In *Proc. of ADMS'14, Fifth International Workshop on Accelerating Data Management Systems Using Modern Processor and Storage Architectures*, August 2014. 64, 65, 66

Hicham Lahlou. Easily accelerating existing Monte-Carlo code, June 2013. 64

Himabindu Lakkaraju, Chiranjib Bhattacharyya, Indrajit Bhattacharya, and Srujana Merugu. Exploiting coherence in reviews for discovering latent facets and associated sentiments. In *Proc. of SIAM International Conference on Data Mining*, April 2011. 35

Laks V.S. Lakshmanan, Jian Pei, and Yan Zhao. Qc-trees: An efficient summary structure for semantic olap. In *Proc. of SIGMOD'03*, June 2003. DOI: 10.1145/872757.872768. 53

Thomas K. Landauer and Susan Dumais. Latent semantic analysis. *Scholarpedia*, 3(11):4356, 2008. 35

Thomas K. Landauer, Peter W. Foltz, and Darrell Laham. An introduction to latent semantic analysis. *Discourse Processes*, 25:259–298, 1998. DOI: 10.1080/01638539809545028. 35

Quoc Le, Marc'Aurelio Ranzato, Rajat Monga, Matthieu Devin, Kai Chen, Greg Corrado, Jeff Dean, and Andrew Ng. Building high-level features using large scale unsupervised learning. In *International Conference in Machine Learning*, 2012. DOI: 10.1109/ICASSP.2013.6639343. 61, 75

Christian Leber, Benjamin Geib, and Heiner Litz. High frequency trading acceleration using fpgas. In *Proc. 2011 International Conference on Field Programmable Logic and Applications*, September 2011. DOI: 10.1109/FPL.2011.64. 63, 64

Y. LeCun and Y. Bengio. Convolution networks for images, speech, and time-series. In M. A. Arbib, editor, *The Handbook of Brain Theory and Neural Networks*, 1995. 27

Y. LeCun, L. Bottou, Y. Bengio, and P. Haffner. Gradient-based learning applied to document recognition. *Proc. of the IEEE*, 86(11):2278–2324, 1998. DOI: 10.1109/5.726791. 27

Daniel Lee and Sabastian Seung. Learning the parts of objects by non-negative matrix factorization. *Nature*, 401:788–791, 1999. DOI: 10.1038/44565. 22, 23, 35

Daniel Lee and Sabastian Seung. Algorithms for non-negative matrix factorization. *Adv. Neural Info. Proc. Systems*, 13:556–562, 2001. 35

Thomas Legler, Wolfgang Lehner, and Andrew Ross. Data Mining with the SAP NetWeaver BI Accelerator, September 2006. 20

Jure Leskovec, Anand Rajaraman, and Jeffrey Ullman. *Mining Massive Datasets*. Cambridge University Press, 2014. 2nd ed., v2.1. 9, 46, 53

T.-S. Lim, W.-Y. Loh, and Y.-S. Shih. A comparison of prediction accuracy, complexity, and training time of thirty-three old and new classification algorithms. *Machine Learning*, 40: 203–229, 2000. DOI: 10.1023/A:1007608224229. 29, 30

G. Linden, B. Smith, and J. York. Amazon.com recommendations: Item-toitem collaborative filtering. *IEEE Internet Computing*, 7(1):76–80, 2003. DOI: 10.1109/MIC.2003.1167344. 23

Daofu Liu, Tianshi Chen, Shaoli Liu, Jinhong Zhou, Shengyuan Zhou, Olivier Temam, Xiaobing Feng, Xuehai Zhou, and Yunji Chen. Pudiannao: A polyvalent machine learning accelerator. In *Proc. of the 20th ACM International Conference on Architectural Support for Programming Languages and Operating Systems (ASPLOS'15)*, March 2015. DOI: 10.1145/2694344.2694358. 62, 76

Ting Liu, Andrew W. Moore, Alexander Gray, and Ke Yang. An investigation of practical approximate nearest neighbor algorithms. In *Proc. of Neural Information Processing Systems, NIPS 2004*, 2004. 18

Ting Liu, Andrew W. Moore, and Alexander Gray. New algorithms for efficient high-dimensional non-parametric classification. *Journal of Machine Learning Research*, 7:1135–1158, 2006. 18

Ting Liu, Charles Rosenberg, and Henry A. Rowley. Clustering billions of images with large scale nearest neighbor search. In *IEEE Workshop on Applications of Computer Vision*, 2007. DOI: 10.1109/WACV.2007.18. 18

Huma Lodhi, Craig Saunders, John Shawe-Taylor, Nello Cristianini, and Chris Watkins. Text classification using string kernels. *Journal of Machine Learning Research*, 2:419–444, 2002. DOI: 10.1162/153244302760200687. 25, 34, 35

W.-Y. Loh and Y.-S. Shih. Split selection methods for classification trees. *Statistica Sinica*, 7: 815–840, 1997. 29, 30

Manoj Lohatepanont and Cynthia Barnhart. Airline Schedule Planning: Integrated Models and Algorithms for Schedule Design and Fleet Assignment. *Transportation Science*, 38(1), February 2004. DOI: 10.1287/trsc.1030.0026. 9

Gabriele Lohmann, Daniel S. Margulies, Annette Horstmann, Burkhard Pleger, Joeran Lepsien, Dirk Goldhahn, Haiko Schloegl, Michael Stumvoll, Arno Villringer, and Robert Turner. Eigenvector centrality mapping for analyzing connectivity patterns in fmri data of the human brain. *PLoS ONE*, 5, 04 2010. DOI: 10.1371/journal.pone.0010232. 46

Gaelle Loosli and Stephane Canu. Comments on the "Core Vector Machines: Fast SVM Training on Very Large Data Sets". *Journal of Machine Learning Research*, 8:291–301, 2007. 25

Jorg Lotze, Paul Sutton, and Hicham Lahlou. Many-core accelerated libor swaption portfolio pricing. *SC Companion*, pages 1185–1192, 2012. DOI: 10.1109/SC.Companion.2012.143. 64

Xiaoyi Lu, Md. Wasi ur Rahman, Nusrat Islam, Dipti Shankar, and Dhabaleswar K. Panda. Accelerating spark with rdma for big data processing: Early experiences.   In *Proc. of 2014 IEEE 22nd Annual Symposium on High-Performance Interconnects*, August 2014. DOI: 10.1109/HOTI.2014.15. 70, 76

Wayne Luk. Accelerating financial computation, May 2013. HPC Finance Conference and Training Event. 63

J. MacQueen. Some methods for classification and analysis of multivariate observations. *Proc. of the Fifth Berkeley Symposium on Mathematical Statistics and Probability*, 1(14):281–297, 1967. 14, 16

Mark Madsen. Advanced Analytics: an overview, 2009. Third Nature, Inc. 9

Michael W. Mahoney. Machine learning and linear algebra for large information graphs. 67

Jim Maltby. uRiKA: A high-performance multi-threaded in-memory graph database, September 2012. Keynote presentation at the ADMS'12 Workshop. 67, 68

Stefan Manegold, Peter A. Boncz, and Martin L. Kersten. Optimizing database architecture for the new bottleneck: Memory access. *The VLDB Journal*, 9(3), December 2000. DOI: 10.1007/s007780000031. 65, 66

Christopher D. Manning, Prabhakar Raghavan, and Hinrich Schutze. *Introduction to Information Retrieval*. Cambridge University Press, 2009. Online edition. 35

James Manyika, Michael Chui, Brad Brown, Jacques Bughin, Richard Dobbs, Charles Roxburgh, and Angela Hung Byers. Big data: The next frontier for innovation, competition, and productivity, May 2011. McKinsey Global Institute. 9

George Marsaglia and Arif Zaman. A new class of random number generators. *The Annals of Applied Probability*, 1(3):462–480, 1991. DOI: 10.1214/aoap/1177005878. 37

Roy Marsten, Radhika Subramanian, Matthew Saltzeman, Irvin Lustig, and David Shanno. Interior point methods for linear programming: Just call newton, lagrange, and fiacco and mccormick. *Interfaces, The Practice of Mathematical Programming*, 20(4), July-August 1990. DOI: 10.1287/inte.20.4.105. 41

M. Matsumoto and T. Nishimura. Mersenne twister: A 623-dimensionally equidistributed uniform pseudorandom number generator. *ACM Transactions on Modeling and Computer Simulation*, 8(1):3–30, January 1998. DOI: 10.1145/272991.272995. 37, 63, 64

M. Matsumoto and T. Nishimura. *Dynamic Creation of Pseudorandom Number Generator*, pages 56–69. Springer, 2000. DOI: 10.1007/978-3-642-59657-5_3. 37

Andrew McCallum and Kamal Nigam. A comparison of event models for naive bayes text classi-fication. In *AAAI/ICML-98 Workshop on Learning for Text Categorization*, pages 41–48, 1998. Technical Report WS-98-05. 35

Adam McLaughlin and David A. Bader. Scalable and high performance betweenness centrality on the gpu. In *Proc. of the 2014 ACM/IEEE Conference on Supercomputing, SC'14*, November 2014. DOI: 10.1109/SC.2014.52. 67, 68

S. Mehrotra. On the implementation of a primal-dual interior point method. *SIAM Journal Optimization*, 2:575–601, 1992. DOI: 10.1137/0802028. 41

P. Melville, R. J. Mooney, and R. Nagarajan. Content-boosted collaborative filtering for improved recommendations. In *Proc. of the Eighteenth National Conference on Artificial Intelligence (AAAI-02), Edmonton, Alberta*, pages 187–192, 2002. 23

Prem Melville and Vikas Sidhwani. Recommender systems. In *Encyclopedia of Machine Learning*, 2010. 21

P. Merolla, J. Arthur, F. Akopyan, N. Imam, R. Manohar, and D. S. Modha. A digital neurosy-naptic core using embedded crossbar memory with 45pj per spike in 45nm. In *Proc. Custom Integrated Circuits Conference (CICC)*, September 2011. DOI: 10.1109/CICC.2011.6055294. 62, 76

Duane Merrill and Andrew Grimshaw. High performance and scalable radix sorting: A case study of implementing dynamic parallelism for GPU computing. *Parallel Processing Letters*, 21 (02):245–272, 2011. DOI: 10.1142/S0129626411000187. 72, 76

Duane Merrill, Michael Garland, and Andrew Grimshaw. Scalable gpu graph traversal. In *Proc. 17th ACM SIGPLAN Symposium on Principles and Practice of Parallel Programming, PPoPP 2012*, February 2012. DOI: 10.1145/2370036.2145832. 67, 68

N. Metropolis, A. W. Rosenbluth, M. N. Rosenbluth, A. H. Teller, and E. Teller. Equations of state calculations by fast computing machines. *Journal of Chemical Physics*, 21(6):1087–1092, 1953. DOI: 10.1063/1.1699114. 37

Nicholas Metropolis and Stanislav Ulam. The monte carlo method. *Journal of the American Statistical Association*, 44(247):335–341, September 1949. DOI: 10.1080/01621459.1949.10483310. 36, 37

S. Micheloyannis, E. Pachou, C. Stam, M. Breakspear, P. Bitsios, M. Vourkas, S. Erimaki, and M. Zervakis. Small-world networks and disturbed functional connectivity in schizophrenia. *Schizophrenia Research*, 87(1-3):60–66, 2006. DOI: 10.1016/j.schres.2006.06.028. 46

Microsoft Corp. Microsoft SQL Server. 20, 44

Microsoft Developer Network. MDX overview. 43, 44

Tomas Mikolov, Kai Chen, Greg Corrado, and Jeffrey Dean. Efficient estimation of word representations in vector space. *CoRR*, abs/1301.3781, 2013. 34

R. Milo, S. Shen-Orr, S. Itzkovitz, N. Kashtan, D. Chklovskii, and U. Alon. Network motifs: Simple building blocks of complex networks. *Science*, 298(5594):824–827, October 2002. DOI: 10.1126/science.298.5594.824. 44, 46

Asit K. Mishra, Rajkishore Barik, and Somnath Paul. iACT: A Software-Hardware Framework For Understanding the Scope of Approximate Computing. In *Proc. of WACAS 2014*, 2014. 73, 76

Abhishek Mitra, Marcos R. Vieira, Petko Bakalov, Walid Najjar, and Vassilis J. Tsotras. Boosting xml filtering with a scalable fpga-based architecture. In *Proc. of the 4th Biennial Conference on Innovative Data Systems Research (CIDR)*, January 2009a. 58

Abhishek Mitra, Marcos R. Vieira, Petko Bakalov, Vassilis J. Tsotras, and Walid A. Najjar. Boosting XML filtering through a scalable FPGA-based architecture. In *Fourth Biennial Conference on Innovative Data Systems Research*, CIDR'09, Asilomar, CA, USA, 2009b. www.cidrdb.org. 72, 76

Andrew Moore. A-star heuristic search, 2011a. http://www.autonlab.org/tutorials/astar.html. 40, 41

Andrew Moore. Constraint satisfaction algorithms, with applications in computer vision and scheduling, 2011b. http://www.autonlab.org/tutorials/constraint.html. 40, 41

Andrew W. Moore. The anchors hierarchy: Using the triangle inequality to survive high-dimensional data. In *Proc. of 12th Conference on Uncertainty in Artificial Intelligence*, 2000. 18

Gareth W. Morris, David B. Thomas, and Wayne Luk. FPGA accelerated low-latency market data feed processing. In *Proc. of the 2009 17th IEEE Symposium on High Performance Interconnects*, HOTI '09, pages 83–89, Washington, DC, USA, 2009. IEEE Computer Society. DOI: 10.1109/HOTI.2009.17. 64

David M. Mount and Sunil Arya. ANN: A Library for Approximate Nearest Neighbor Searching, January 2010. www.cs.umd.edu/~mount/ANN. 18

Sreerama K. Murthy. Automatic construction of decision trees from data: A multi-disciplinary survey. *Data Mining and Knowledge Discovery*, 2:345–389, 1998. DOI: 10.1023/A:1009744630224. 28, 30

Teemu Mutanen. Customer churn analysis—a case study. Technical Report VTT-R-01184-06, Technical Research Centre of Finland, 2006. 7

Ravi Nair. Models for energy-efficient approximate computing. In *Proc. of the 16th ACM/IEEE International Symposium on Low Power Electronics and Design*, ISLPED '10, pages 359–360, 2010. DOI: 10.1145/1840845.1840921. 76

Ravi Nair. Big data needs approximate computing: Technical perspective. *Commun. ACM*, 58 (1):104–104, December 2014. DOI: 10.1145/2688072. 73, 76

Mohammadreza Najafi, Mohammad Sadoghi, and Hans-Arno Jacobsen. Flexible query processor on fpgas. *Proc. VLDB Endow.*, 6(12):1310–1313, August 2013. DOI: 10.14778/2536274.2536303. 76

Amit A. Nanavati, Siva Gurumurthy, Gautam Das, Dipanjan Chakraborty, Koustuv Dasgupta, Sougata Mukherjea, and Anupam Joshi. On the structural properties of massive telecom call graphs: Findings and implications. In *Proc. of the Conference on Information and Knowledge Management (CIKM'06)*, pages 435–444, November 2006. 7, 45, 46

Network-Enabled Optimization Systems Wiki. Nonlinear programming FAQ. http://www.neos-guide.org/NEOS/index.php/Nonlinear_Programming_FAQ. DOI: 10.1145/1183614.1183678. 41

Arnold Neumaier. Complete search in continuous global optimization and constraint satisfaction. *Acta Numerica*, 2004. DOI: 10.1017/S0962492904000194. 41

Padraic G. Neville. Decision tress for predictive modeling, 1999. 30

M. E. J. Newman. *Networks: An Introduction*. Oxford University Press, March 2010. 44, 45, 46

Andrew Ng. Deep learning, March 2015. Nvidia Global Technology Conference Keynote. 59, 61, 76

E. W. T. Ngai, L. Xiu, and D. C. K. Chau. Application of data mining techniques in customer relationship management: A literature review and classification. *Expert Systems with Applications*, 36:2592–2602, 2009. DOI: 10.1016/j.eswa.2008.02.021. 9

Nils Nilsson. *Principles of Artificial Intelligence*. Morgan Kaufmann, 1980. DOI: 10.1007/978-3-662-09438-9. 40, 41

NIPS 2009 Workshop. Large-Scale Machine Learning: Parallelism and Massive Datasets, December 2009. 9

Robert Nisbet, John Elder, and Gary Miner. *Handbook of Statistical Analysis and Data Mining Applications*. Academic Press, 2009. 9, 16, 18, 30, 33

Feng Niu, Benjamin Recht, Christopher Re, and Stephen J. Wright. HOGWILD!: A Lock-Free Approach to Parallelizing Stochastic Gradient Descent. In *Advances in Neural Information Processing Systems*, 2011. 70

William S. Noble. What is a support vector machine? *Nature Biotechnology*, 24(12), December 2006. DOI: 10.1038/nbt1206-1565. 23, 25

S. M. Omohundro. Efficient algorithms with neural network behaviour. *Journal of Complex Systems*, 1(2):273–347, 1987. 17, 18

S. M. Omohundro. Five balltree construction algorithms. Technical Report TR-89-063, ICSI, 1989. 18

S. M. Omohundro. Bumptrees for efficient function, constraint, and classification learning. *Advances in Neural Information Processing Systems*, 3, 1991. 18

Oracle Corp. Oracle 11g Database OLAP, a. `www.oracle.com/technetwork/database/options/olap/index.html`. 44

Oracle Corp. Oracle Data Miner 11g Release 2, b. 14, 16, 20, 25, 30, 35

Jian Ouyang, Shiding Lin, Wei Qi, Yong Wang, Bo Yu, and Song Jiang. Sda: Software-defined accelerator for large-scale dnn system, August 2014. 61, 62, 76

Kalin Ovtcharov, Olatunji Ruwase, Joo-Young Kim, Jeremy Fowers, Karin Strauss, and Eric S. Chung. Accelerating deep convolutional neural networks using specialized hardware, February 2015. 61, 62, 76

Lawrence Page, Sergey Brin, Rajeev Motwani, and Terry Winograd. The pagerank citation ranking: Bringing order to the web. Technical Report 1999-66, Stanford InfoLab, November 1999. Previous number = SIDL-WP-1999-0120. 46

Bo Pang, Lillian Lee, and Shivakumar Vaithyanathan. Thumbs up? sentiment classification using machine learning techniques. In *Processings of EMNLP*, pages 79–86, 2002. DOI: 10.3115/1118693.1118704. 35

Christos H. Papadimitriou and Kenneth Steiglitz. *Combinatorial Optimization: Algorithms and Complexity*. Prentice-Hall, 1998. 39, 41

Grigorios Papamanousakis, Jinzhe Yang, and Grzegorz Kozikowski. Potential future exposure and collateral modelling of the trading book using nvidia gpus, March 2015. Nvidia Global Technology Conference. 64

M. J. Pazzani. A framework for collaborative, content-based and demographic filtering. *Artificial Intelligence Review*, 13(5-6):313–331, 1999. DOI: 10.1023/A:1006544522159. 23

M. J. Pazzani and D. Billsus. Learning and revising user profiles: The identification of interesting web sites. *Machine Learning*, 27(3):313–331, 1997. DOI: 10.1023/A:1007369909943. 23

Jeffrey Pennington, Richard Socher, and Christopher D. Manning. Glove: Global vectors for word representation. In *Proc. of Conference on Empirical Methods in Natural Language Processing (EMNLP 2014)*, October 2014. DOI: 10.3115/v1/D14-1162. 34

Francisco Pereira, Tom Mitchell, and Matthew Botvinick. Machine learning classifiers and fMRI: a tutorial overview. *NeuroImage*, 45(1):S199–S209, 2009. Mathematics in Brain Imaging. DOI: 10.1016/j.neuroimage.2008.11.007. 23

S. H. Peskov and J. F. Traub. Faster evaluation of financial derivatives. *Journal of Portfolio Management*, 22(1):113–120, 1995. DOI: 10.3905/jpm.1995.409541. 37

Phi-Hung Pham, Darko Jelaca, Clement Farabet, Berin Martini, Yann LeCun, and Eugenio Culurciello. Neuflow: Dataflow vision processing system-on-a-chip. In *Proc. IEEE International Midwest Symposium on Circuits and systems, IEEE MWSCAS*, 2012. DOI: 10.1109/MWSCAS.2012.6292202. 61, 62, 72, 76

Gregory Piatetsky-Shapiro. Kdnuggets, 2011. www.kdnuggets.com. 9

Yevgeniy Podolyan and George Karypis. Common pharmacophore identification using frequent clique detection algorithm. *Journal of Chemical Information and Modeling*, 2008. DOI: 10.1021/ci8002478. 46

Mason A. Porter, Jukka-Pekka Onnela, and Peter J. Mucha. Communities in networks. *Notices of the AMS*, 56(9), October 2009. 46

Andrew Putnam, Adrian M. Caulfield, Eric S. Chung, Derek Chiou, Kypros Constantinides, John Demme, Hadi Esmaeilzadeh, Jeremy Fowers, Gopi Prashanth Gopal, Jan Gray, Michael Haselman, Scott Hauck, Stephen Heil, Amir Hormati, Joo-Young Kim, Sitaram Lanka, James Larus, Eric Peterson, Simon Pope, Aaron Smith, Jason Thong, Phillip Yi Xiao, and Doug Burger. A reconfigurable fabric for accelerating large-scale datacenter services. In *Proc. of the 41st Annual International Symposium on Computer Architecuture*, ISCA '14, pages 13–24, Piscataway, NJ, USA, 2014. IEEE Press. http://dl.acm.org/citation.cfm?id=2665671.2665678. DOI: 10.1145/2678373.2665678. 72, 76

J. Ross Quinlan. Induction of decision trees. *Machine Learning*, 1(1):81–106, March 1986. DOI: 10.1023/A:1022643204877. 29, 30

J. Ross Quinlan. *C4.5: Programs for Machine Learning*. Morgan Kaufmann Publishers, 1993. 29, 30

Vijayshankar Raman, Gopi Attaluri, Ronald Barber, Naresh Chainani, David Kalmuk, Vincent KulandaiSamy, Jens Leenstra, Sam Lightstone, Shaorong Liu, Guy M. Lohman, Tim Malkemus, Rene Mueller, Ippokratis Pandis, Berni Schiefer, David Sharpe, Richard Sidle, Adam Storm, and Liping Zhang. Db2 with blu acceleration: So much more than just a column store. *Proc. VLDB Endow.*, 6(11), August 2013. DOI: 10.14778/2536222.2536233. 65, 66

A. Ravishankar Rao, Rahul Garg, and Guillermo A. Cecchi. A Spatio-Temporal Support Vector Machine Searchlight for fMRI Analysis. In *Proc. of 2011 IEEE International Symposium on Biomedial Imaging: From Nano to Macro*, March-April 2011. DOI: 10.1109/ISBI.2011.5872575. 23, 25

Rapid-i. RapidMiner Data and Text Mining. `rapid-i.com/`. 14, 16, 20, 25, 28, 30, 33, 35

Karl Rexer. 2013 data miner survey highlights. In *Procs. of the Predictive Analytics World*, October 2013. 9, 13

Yossi Richter, Elad Yom-Tov, and Noam Slonim. Predicting customer churn in mobile networks through analysis of social groups. In *Procs. of the SIAM International Conference on Data Mining, SDM 2010*, pages 732–741, 2010. 2, 9, 45, 46

Erik Riedel, Garth Gibson, and Christos Faloutsos. Active storage for large-scale data mining and multimedia. In *Proc. of the 24th VLDB Conference*, September 1998. 73

M. Roesch. Snort- lightweight intrusion detection for networks. In *Proc. LISA99, the 13th Systems Administration Conference*, 1999. 57, 58

Indranil Roy and Srinivas Aluru. Finding motifs in biological sequences using the micron automata processor. In *Proc. 2014 IEEE 28th International Parallel and Distributed Processing Symposium*, May 2014. DOI: 10.1109/IPDPS.2014.51. 58

Mohammad Sadoghi, Martin Labrecque, Harsh Singh, Warren Shum, and Hans-Arno Jacobsen. Efficient event processing through reconfigurable hardware for algorithmic trading. *Proc. of the VLDB Endowment*, 3(2), 2010. DOI: 10.14778/1920841.1921029. 64

M. Sahami, S. Dumais, D. Heckerman, and E. Horvitz. A bayesian approach to filtering junk e-mail. In *AAAI/ICML-98 Workshop on Learning for Text Categorization*, 1998. 35

M. Saito and M. Matsumoto. *SIMD-oriented Fast Mersenne Twister: a 128-bit Pseudorandom Number Generator*, pages 607–622. Springer, 2008. DOI: 10.1007/978-3-540-74496-2_36. 37

Valentina Salapura, Tejas Karkhanis, Priya Nagpurkar, and Jose Moreira. Accelerating business analytics applications. In *Proc. IEEE 18th International Symposium on High Performance Computer Architecture (HPCA)*, February 2012. DOI: 10.1109/HPCA.2012.6169044. 58

SAP Inc. In-memory Computing: Better Insight Faster with the SAP In-memory Appliance (SAP HANA), December 2010. 44

Warren S. Sarle. Neural network FAQ, periodic posting to the usenet newsgroup comp.ai.neural-nets, 2002. `ftp://ftp.sas.com/pub/neural/FAQ.html`. 25, 28

SAS Institute Inc. SAS Analytics. 14, 16, 20, 25, 28, 30, 33

Ashok Savasere, Edward Omiecinski, and Shamkant Navathe. An efficient algorithm for mining association rules in large databases. In *Proc. of the 21st VLDB Conference*, pages 432–444, 1995. 20

Shlomo S. Sawilowsky. You think you've got trivials? *Journal of Modern Applied Statistical Methods*, 2(1), May 2003. 36, 37

Satu Elisa Schaeffer. Graph clustering. *Computer Science Review I*, pages 27–64, 2007. DOI: 10.1016/j.cosrev.2007.05.001. 46

J. Ben Schafer, Joseph Konstan, and John Riedl. Recommender systems in e-commerce. In *Proc. of E-COMMERCE99*, 1999. DOI: 10.1145/336992.337035. 21

Benjamin Schlegel, Thomas Willhalm, and Wolfgang Lehner. Fast sorted-set intersection using simd instructions. In *Proc. of ADMS'11, Second International Workshop on Accelerating Data Management Systems Using Modern Processor and Storage Architectures*, August 2013. 58, 65, 66

J. Schmidhuber. Deep learning in neural networks: An overview. *CoRR*, abs/1404.7828, 2014. DOI: 10.1016/j.neunet.2014.09.003. 27, 61

Alexander Schrijver. *Theory of Linear and Integer Programming*. John Wiley & Sons, 1998. 39, 41

Alexander Schrijver. *Combinatorial Optimization: Polyhedra and Efficiency*. Springer, 2002. 41

Science Special Issue. Dealing with data. *Science*, 331(6018), February 2011. 9

Fabrizio Sebastiani. Machine learning in automated text categorization. *ACM Computing Surveys*, 34:1–47, 2002. DOI: 10.1145/505282.505283. 34

Jae-sun Seo, Bernard Brezzo, Yong Liu, Benjamin D. Parker, Steven K. Esser, Robert K. Montoye, Bipin Rajendran, José A. Tierno, Leland Chang, Dharmendra S. Modha, and Daniel J. Friedman. A 45nm CMOS neuromorphic chip with a scalable architecture for learning in networks of spiking neurons. In *2011 IEEE Custom Integrated Circuits Conference, CICC 2011, San Jose, CA, USA, Sept. 19-21, 2011*, pages 1–4, 2011. DOI: 10.1109/CICC.2011.6055293. 61, 62, 72, 76

Martin Sewell. Support vector machines: Financial applications. `http://www.svms.org/finance/`. 23, 25

Guangyu Shi, Min Li, and Mikko Lipasti. Accelerating search and recognition workloads with sse 4.2 string and text processing instructions. In *Proc. ISPASS'11*, 2011. DOI: 10.1109/ISPASS.2011.5762731. 58

Y-.S. Shih. *QUEST User Manual*, April 2004. 28, 30

Galit Shmueli, Nitin R. Patel, and Peter C. Bruce. *Data Mining for Business Intelligence: Concepts, Techniques, and Applications in Microsoft Office Excel with XLMiner*. John Wiley & Sons Inc., 2010. 2nd ed. 9, 12, 14, 16, 30, 33

Robert H. Shumway and David S. Stoffer. *Time Series Analysis and Its Applications: With R Examples (3rd ed.)*. Springer Texts in Statistics, 2010. DOI: 10.1007/978-1-4419-7865-3. 31, 33

Vishal Sikka, Franz Färber, Anil Goel, and Wolfgang Lehner. Sap hana: The evolution from a modern main-memory data platform to an enterprise application platform. *Proc. VLDB Endow.*, 6(11), August 2013. DOI: 10.14778/2536222.2536251. 65, 66

Vikas Sindhwani, Amol Ghoting, Edison Ting, and Richard Lawrence. Extracting insights from social media with large-scale matrix approximations. *IBM Journal of Research and Development*, 55(5):9:1–9:13, Sept-Oct 2011. DOI: 10.1147/JRD.2011.2163281. 9

Gordon K. Smyth. Nonlinear regression. *Encyclopedia of Environmetrics*, 3:1405–1411, 2002. 14

William S. Song, Jeremy Kepner, Vitaliy Gleyzer, Huy T. Nguyen, and Joshua I. Kramer. Novel graph processor architecture. *Lincoln Laboratory Journal*, 20(1), 2013. 68, 72, 76

James C. Spall. *Introduction to Stochastic Search and Optimization: Estimation, Simulation, and Control*. Wiley, 2003. DOI: 10.1002/0471722138. 41

Splunk Inc. *Splunk Tutorial*, 2011. `www.splunk.com`. 9

C. Stam, B. Jones, G. Nolte, M. Breakspear, and P. Scheltens. Small-world networks and functional connectivity in Alzheimer's disease, 2006. Celebral Cortex. 46

David T. Stanton, Timothy W. Morris, Siddhartha Roychoudhury, and Christian N. Parker. Application of nearest-neighbor and cluster analyses in pharmaceutical lead discovery. *J. Chem. Inf. Comput. Sci.*, 39(1):21–27, 1999. DOI: 10.1021/ci9801015. 16, 18

StatSoft Inc. StatSoft Electronic Statistics Textbook, 2010. `http://www.statsoft.com/textbook`. 9, 14, 31, 32, 33

Christos Stergiou and Dimitrios Siganos. *Neural Networks*. 25, 28

Alexander Strehl, Joydeep Ghosh, and Raymond Mooney. Impact of similarity measures on webpage clustering. In *AAAI-2000: Workshop of Artificial Intelligence for Web Search*, pages 58–64, July 2000. 34

Peter Strohm. Multi-dimensional, in-gpu-memory databases: Streaming conditional calculations in big data sets, 2015. Nvidia Global Developer Conference. 65

Ari Studnitzer and Oskar Mencer. Going to the wire: The next generation financial risk management platform. In *Proc. Hot Chips: A Symposium on High Performance Chips (HC25)*, August 2013. 64

X. Su and T. M. Khoshgoftaar. A survey of collaborative filetring techniques. In *Advances in Artificial Intelligence*, pages 1–20, 2009. DOI: 10.1155/2009/421425. 23

Ambika Suman. Automated face recognition, applications within law enforcement: Market and technology review, October 2006. 9

Christian Szegedy, Wei Liu, Yangqing Jia, Pierre Sermanet, Scott Reed, Dragomir Anguelov, Dumitru Erhan, Vincent Vanhoucke, and Andrew Rabinovich. Going deeper with convolutions. *CoRR*, abs/1409.4842, 2014. 59, 61

Teradata Inc. Teradata Database 13.10. 44

The Economist. Algorithms: Business by numbers, 2007. Print Edition, September 13, 2007. 9

The Economist. Data, data everywhere, 2010. Print Edition, February 25, 2010. 1

The R Foundation. The R Foundation for Statistical Computing. www.r-project.org. 14, 16, 20, 25, 28, 30, 33, 46

The StatSoft Inc. STATISTICA version 10. www.statsoft.com. 14, 16, 20, 25, 28, 30, 33, 35

David Thomas, Lee Howes, and Wayne Luk. A comparison of cpus, gpus, fpgas, and massively parallel processor arrays for random number generation. In *Proc. of FPGA'09*, February 2009. DOI: 10.1145/1508128.1508139. 63, 64

Henry Thompson. Parallel parsers for context-free grammars: Two actual implementations compared. *Parallel Natural Language Processing*, pages 168–187, 1994. 57, 58

Michael J. Todd. The many facets of linear programming. *Mathematical Programming*, 91(3): 417–436, 2002. DOI: 10.1007/s101070100261. 39, 41

Ivor W. Tsang, James T. Kwok, and Pat-Ming Cheung. Core Vector Machines: Fast SVM Training on Very Large Data Sets. *Journal of Machine Learning Research*, 6:363–392, 2005. 25

Antonino Tumeo, Oreste Villa, and Donatella Sciuto. Efficient pattern matching on GPUs for intrusion detection systems. In *Proc. of the 7th ACM international conference on Computing frontiers*, CF'10, pages 87–88, Bertinoro, Italy, 2010. ACM. DOI: 10.1145/1787275.1787296. 76

Jeffrey K. Uhlmann. Satisfying general proximity/similarity queries with metric trees. *Information Processing Letters*, 40:175–179, 1991. DOI: 10.1016/0020-0190(91)90074-R. 17, 18

Mustafa Uysal, Anurag Acharya, and Joel H. Saltz. Evaluation of active disks for decision support databases. In *Proc. of Conference on High Performanc Computer Architecture*, 2000. DOI: 10.1109/HPCA.2000.824363. 73

Ewout van den Berg, Daniel Brand, Leonid Rachevsky, Rajesh Bordawekar, and Bhuvana Rambhadran. Accelerating deep convolution neural networks for large-scale speech tasks using gpus, March 2015. *Nvidia Global Technology Conference*. 60, 61

M. van den Heuvel, C. Stam, H. Boersma, and H. H. Pol. Small-world and scale-free organization of voxel-based resting-state functional connectivity in the human brain. *NeuroImage*, 43 (3):528–539, 2008. DOI: 10.1016/j.neuroimage.2008.08.010. 44, 46

Vladimir Vapnik. *The Nature of Statistical Learning Theory*. Springer-Verlag, New York, 1995. DOI: 10.1007/978-1-4757-3264-1. 23, 25

Vladimir Vapnik. *Statistical Learning Theory*. John Wiley, and Sons, Inc., New York, 1998. 23, 25

Nicolas Vasilache, Jeff Johnson, Michaël Mathieu, Soumith Chintala, Serkan Piantino, and Yann LeCun. Fast convolutional nets with fbfft: A GPU performance evaluation. *CoRR*, abs/1412.7580, 2014. 59

Vijay Vazirani. *Approximation Algorithms*. Springer, 2003. DOI: 10.1007/978-3-662-04565-7. 41

Vertica Systems Inc. The Vertica Analytic Database Technical Overview White Paper, March 2010. 44

Basant Vinaik and Rahoul Puri. Oracle's sonoma processor: Advanced low-cost sparc processor for enterprise workloads, August 2015. *Hot Chips 27*. 66

Amy Wang, Jan Treibig, Bob Blainey, Peng Wu, Yaoqing Gao, Barnaby Dalton, Danny Gupta, Fahham Khan, Neil Bartlett, Lior Velichover, James Sedgwick, and Louis Ly. Optimizing ibm algorithmics' mark-to-future aggregation engine for real-time counterparty credit risk scoring. In *Proc. of the 6th Workshop on High Performance Computational Finance*, WHPCF '13, 2013. DOI: 10.1145/2535557.2535567. 64

Huayong Wang, Henrique Andrade, Buğra Gedik, and Kun-Lung Wu. A code generation approach for auto-vectorization in the spade compiler. In *Proc. of the 22nd International Conference on Languages and Compilers for Parallel Computing*, LCPC'09, 2010. DOI: 10.1007/978-3-642-13374-9_26. 65

Jingdong Wang, Heng Tao Shen, Jingkuan Song, and Jianqiu Ji. Hashing for similarity search: A survey, August 2014. arXiv:1408.29227v1. 18

Yangzihao Wang and John Owens. Large-scale graph processing algorithms on the gpu. Technical Report, Electrical and Computer Engineering Department, UC Davis, January 2013. 67

Yangzihao Wang, Andrew Davidson, Yuechao Pan, Yuduo Wu, Andy Riffel, and John D. Owens. Gunrock: A high-performance graph processing library on the gpu. In *Proc. 20th ACM SIGPLAN Symposium on Principles and Practice of Parallel Programming*, PPoPP 2015, pages 265–266, February 2015. DOI: 10.1145/2688500.2688538. 67, 68

Duncan J. Watts and Steven H. Strogatz. Collective dynamics of 'small-world' networks. *Nature*, 393:440–442, June 1998. DOI: 10.1038/30918. 45, 46

Norbert Wehn. Hardware accelerators for financial mathematics- methodology, results, and benchmarks. In *Proc. 11th International Forum on Embedded MPSoC and Multicore*, July 2011. 62, 64

Eric W. Weisstein. Predictor-corrector methods. In *MathWorld- A Wolfram Web Resource*, http://mathworld.wolfram.com/Predictor-CorrectorMethods.html, 2011. 41

S. Weston, J-T. Marin, J. Spooner, O. Pell, and O. Mencer. Accelerating the computation of portfolios of tranched credit derivatives. In *Proc. IEEE Workshop on High Performance Computational Finance*, November 2010. DOI: 10.1109/WHPCF.2010.5671822. 64

S. Weston, J. Spooner, S. Racaniere, and O. Mencer. Rapid computation of value and risk for derivatives portfolios. *Concurrency and Computation: Practice and Experience*, July 2011. DOI: 10.1002/cpe.1778. 64

Bernard Widrow, David E. Rumelhart, and Michael A. Lehr. Neural networks: applications in industry, business and science. *Communications of the ACM*, 37:93–105, March 1994. DOI: 10.1145/175247.175257. 25, 28

wiki. Wikipedia. www.wikipedia.org. 18

Thomas Willhalm, Nicolae Popovici, Yazan Boshmaf, Hasso Plattner, Alexander Zeier, and Jan Schaffner. SIMD-scan: ultra fast in-memory table scan using on-chip vector processing units. *Proc. VLDB Endowment*, 2:385–394, August 2009a. DOI: 10.14778/1687627.1687671. 44

Thomas Willhalm, Nicolae Popovici, Yazan Boshmaf, Hasso Plattner, Alexander Zeier, and Jan Schaffner. Simd-scan: Ultra fast in-memory table scan using on-chip vector processing units. *Proc. VLDB Endow.*, 2(1), August 2009b. DOI: 10.14778/1687627.1687671. 65, 66

Haicheng Wu, Gregory Diamos, Tim Sheard, Molham Aref, Sean Baxter, Michael Garland, and Sudhakar Yalamanchili. Red fox: An execution environment for relational query processing on gpus. In *Proc. of Annual IEEE/ACM International Symposium on Code Generation and Optimization*, CGO '14, pages 44:44–44:54, New York, USA, 2014a. ACM. DOI: 10.1145/2581122.2544166. 72, 76

Haicheng Wu, Gregory Diamos, Tim Sheard, Molham Aref, Sean Baxter, Michael Garland, and Sudhakar Yalamanchili. Red fox: An execution environment for relational query processing on gpus. In *Proc. of ADMS'14, Fifth International Workshop on Accelerating Data Management Systems Using Modern Processor and Storage Architectures*, August 2014b. DOI: 10.1145/2544137.2544166. 65

Lisa Wu, Andrea Lottarini, Timothy K. Paine, Martha A. Kim, and Kenneth A. Ross. Q100: The Architecture and Design of a Database Processing Unit. In *Proc. of Conference on Architectural Support for Programming Languages and Operating Systems*, March 2014c. DOI: 10.1145/2541940.2541961. 66, 72

Ren Wu, Shengen Yan, Yi Shan, Qingqing Dang, and Gang Sun. Deep image: Scaling up image recognition. *CoRR*, abs/1501.02876, 2015. 61, 75, 76

Xindong Wu, Vipin Kumar, J. Ross Quinlan, Joydeep Ghosh, Qiang Yang, Hiroshi Motoda, Geoffrey J. McLachlan, Angus Ng, Bing Liu, Philip S. Yu, Zhi-Hua Zhou, Michael Steinbach, Hand David J, and Dan Steinberg. Top 10 algorithms in data mining. *Knowledge Information Systems*, 14:1–37, 2008. DOI: 10.1007/s10115-007-0114-2. 9, 14, 18, 30

xcelerit. Accelerating CVA Computation the Easy Way, March 2013a. 64

xcelerit. Derivative Pricing on Altera's FPGAs, April 2013b. 63, 64

xcelerit. Coding for Xeon Phi the Easy Way, December 2014. xcelerit Blog. 63, 64

Wei Xu, Xin Liu, and Yihong Gong. Document clustering based on non-negative matrix factorization. In *Proc. of SIGIR'03*, August 2003. DOI: 10.1145/860435.860485. 35

Xintian Yang, Srinivasan Parthasarathy, and P. Sadayappan. Fast sparse matrix-vector multiplication on gpus: Implications for graph mining. In *Proc. 37th International Conference on Very Large Data Bases*, August 2011. DOI: 10.14778/1938545.1938548. 68

Youngmin Yi, Chao-Yue Lai, Slav Petrov, and Kurt Keutzer. Efficient parallel cky parsing on gpus. In *Proc. 2011 Conference on Parsing Technologies*, October 2011. 57, 58

Mohammed Zaki, Srinivasan Parthasarathy, Mitsunori Ogihara, and Wei Li. New algorithms for fast discovery of association rules. In *Proc. of the 3rd Int'l Conf. on Knowledge Discovery and Data Mining*, pages 283–286, 1997. DOI: 10.1023/A:1009773317876. 20

Mohammed J. Zaki. Scalable algorithms for association mining. *IEEE Transactions on Knowledge and Data Engineering*, 12(3):372–390, 2000. DOI: 10.1109/69.846291. 20

Matthew Zeiler. Clarifai: Enabling next geneation intelligent applications. Nvidia GTC'14 Presentation. 61

Chen Zhang, Peng Li, Guangyu Sun, Yijin Guan, Bingjun Xiao, and Jason Cong. Optimizing fpga-based accelerator design for deep convolutional neural networks. In *Proc. of 23rd International Symposium on Field-Programmable Gate Arrays (FPGA2015)*, February 2015. DOI: 10.1145/2684746.2689060. 61, 62, 76

Tian Zhang, Raghu Ramakrishnan, and Miron Livny. Birch: An efficient data clustering method for very large databases. In *Proc. of the ACM SIGMOD Conference*, 1996. DOI: 10.1145/235968.233324. 15, 16

Yongpeng Zhang, Frank Mueller, Xiaohui Cui, and Thomas Potok. Gpu-accelerated text mining. In *Proc. EPHAM'09*, March 2009. 58

Ying Zhao and George Karypis. Hierarchical clustering algorithms for document datasets. *Data Mining and Knowledge Discovery*, 10(2):141–168, 2005a. DOI: 10.1007/s10618-005-0361-3. 34

Ying Zhao and George Karypis. Topic-driven clustering for document datasets. In *SIAM International Conference on Data Mining*, pages 358–369, 2005b. DOI: 10.1137/1.9781611972757.32. 34, 35

Jingren Zhou and Kenneth A. Ross. Implementing database operations using simd instructions. In *Proc. of the 2002 ACM SIGMOD International Conference on Management of Data*, SIGMOD '02, pages 145–156, 2002. DOI: 10.1145/564691.564709. 65, 66

Y. Zhou, D. Wilkinson, R. Schreiber, and R. Pan. Large-scale parallel collaborative filtering for the netflix prize. In *Proc. of the 4th International Conference on Algorithmic Aspects in Information and Management*, pages 337–348, 2008. DOI: 10.1007/978-3-540-68880-8_32. 22, 23

Authors' Biographies

RAJESH BORDAWEKAR

Rajesh Bordawekar is a Research Staff Member at the IBM T. J. Watson Research Center. His current interest is exploring software-hardware co-design of analytics workloads. He works at the intersection of high-performance computing, analytics, and data management domains. He has published over 40 technical publications and issued 14 patents. He has also presented tutorials at top conferences including ISCA, ASPLOS, and PPoPP. Recently, he has been investigating how GPUs could be used for accelerating key analytics kernels in text analytics, data management, graph analytics, and deep learning. As part of this work, he collaborates closely with the IBM Power Systems, and various analytics and database product teams.

BOB BLAINEY

Bob Blainey is an IBM Fellow and the chief architect of the IBM CloudLab. Bob has been with IBM for over 25 years, with a consistent focus on deep optimization of software for IBM systems, and now IBM's cloud. He spent many years working on programming languages, compilers, tools, and algorithms for parallelism and for high performance on systems, at the microprocessor, node, and cluster level. Bob is now leading a team focused on the invention of the next generation of IBM's cloud infrastructure, driving leadership efficiency, scale, and performance.

RUCHIR PURI

Ruchir Puri is an IBM Fellow at IBM Thomas J Watson Research Center where he leads research efforts in system design and acceleration. Most recently, he led the design methodology innovations for IBM's Power and zEnterprise microprocessors. Dr. Puri has received numerous accolades including the highest technical position at IBM, the IBM Fellow, which was awarded for his transformational role in microprocessor design methodology. In addition, he has received "Best of IBM" awards in both 2011 and 2012 and IBM Corporate Award from IBM's CEO, and several IBM Outstanding Technical Achievement awards. Dr. Puri is a Fellow of the IEEE, an ACM Distinguished Speaker and IEEE Distinguished Lecturer. He is also a member of IBM Academy of Technology and IBM Fellow leadership team and was appointed an IBM Master Inventor in 2010. Dr. Puri is a recipient of Semiconductor Research Corporation (SRC) outstanding mentor award and has been an adjunct professor in the Department of Electrical Engineering at Columbia University, and was also honored with the John Von-Neumann Chair at Institute of Discrete Mathematics at Bonn University, Germany. Dr. Puri is also a recipient of the 2014 Asian American Engineer of the Year Award.